Communications in Computer and Information Science 2105

Rationale
The CCIS series is devoted to the publication of proceedings of computer science conferences. Its aim is to efficiently disseminate original research results in informatics in printed and electronic form. While the focus is on publication of peer-reviewed full papers presenting mature work, inclusion of reviewed short papers reporting on work in progress is welcome, too. Besides globally relevant meetings with internationally representative program committees guaranteeing a strict peer-reviewing and paper selection process, conferences run by societies or of high regional or national relevance are also considered for publication.

Topics
The topical scope of CCIS spans the entire spectrum of informatics ranging from foundational topics in the theory of computing to information and communications science and technology and a broad variety of interdisciplinary application fields.

Information for Volume Editors and Authors
Publication in CCIS is free of charge. No royalties are paid, however, we offer registered conference participants temporary free access to the online version of the conference proceedings on SpringerLink (http://link.springer.com) by means of an http referrer from the conference website and/or a number of complimentary printed copies, as specified in the official acceptance email of the event.

CCIS proceedings can be published in time for distribution at conferences or as post-proceedings, and delivered in the form of printed books and/or electronically as USBs and/or e-content licenses for accessing proceedings at SpringerLink. Furthermore, CCIS proceedings are included in the CCIS electronic book series hosted in the SpringerLink digital library at http://link.springer.com/bookseries/7899. Conferences publishing in CCIS are allowed to use Online Conference Service (OCS) for managing the whole proceedings lifecycle (from submission and reviewing to preparing for publication) free of charge.

Publication process
The language of publication is exclusively English. Authors publishing in CCIS have to sign the Springer CCIS copyright transfer form, however, they are free to use their material published in CCIS for substantially changed, more elaborate subsequent publications elsewhere. For the preparation of the camera-ready papers/files, authors have to strictly adhere to the Springer CCIS Authors' Instructions and are strongly encouraged to use the CCIS LaTeX style files or templates.

Abstracting/Indexing
CCIS is abstracted/indexed in DBLP, Google Scholar, EI-Compendex, Mathematical Reviews, SCImago, Scopus. CCIS volumes are also submitted for the inclusion in ISI Proceedings.

How to start
To start the evaluation of your proposal for inclusion in the CCIS series, please send an e-mail to ccis@springer.com.

Oleg Gusikhin · Slimane Hammoudi ·
Alfredo Cuzzocrea
Editors

Data Management Technologies and Applications

12th International Conference, DATA 2023
Rome, Italy, July 11–13, 2023
Revised Selected Papers

Springer

Editors
Oleg Gusikhin
Ford Motor Company
Dearborn, MI, USA

Slimane Hammoudi
ESEO, ERIS
Angers Cedex, France

Alfredo Cuzzocrea
University of Calabria
Rende, Italy

ISSN 1865-0929 ISSN 1865-0937 (electronic)
Communications in Computer and Information Science
ISBN 978-3-031-68918-5 ISBN 978-3-031-68919-2 (eBook)
https://doi.org/10.1007/978-3-031-68919-2

This Springer imprint is published by the registered company Springer Nature Switzerland AG
The registered company address is: Gewerbestrasse 11, 6330 Cham, Switzerland

Preface

The present book includes extended and revised versions of a set of selected papers from the 12th International Conference on Data Science, Technology and Applications (DATA 2023), held in Rome, Italy, from 11–13 July 2023.

DATA 2023 received 106 paper submissions from 34 countries, of which six (6%) were included in this book.

The papers were selected by the event chairs and their selection was based on a number of criteria that included the classifications and comments provided by the program committee members, the session chairs' assessment and also the program chairs' global view of all papers included in the technical program. The authors of selected papers were then invited to submit revised and extended versions of their papers having at least 30% innovative material.

The purpose of the International Conference on Data Science, Technology and Applications (DATA) is to bring together researchers, engineers and practitioners interested in databases, big data, data mining, data management, data security and other aspects of information systems and technology involving advanced applications of data.

The papers selected to be included in this book contribute to the understanding of relevant trends of current research on Data Science, Technology and Applications, including: Big Data Applications, Data Analytics, Data Science, NoSQL Databases, Social Data Analytics, Dimensional Modelling, Deep Learning and Big Data, Decision Support Systems, Data Warehouse Management and Data Management for Analytics.

We would like to thank all the authors for their contributions and also all reviewers who have helped to ensure the quality of this publication.

July 2023

Oleg Gusikhin
Slimane Hammoudi
Alfredo Cuzzocrea

Organization

Conference Chair

Alfredo Cuzzocrea	University of Calabria, Italy

Program Co-chairs

Oleg Gusikhin	Ford Motor Company, USA
Slimane Hammoudi	ESEO, ERIS, France

Program Committee

Toshiyuki Amagasa	University of Tsukuba, Japan
Bernd Amann	Sorbonne Université, France
Adnan Amin	Institute of Management Sciences, Pakistan
Theodoros Anagnostopoulos	University of West Attica, Greece
Jean Barddal	Pontifícia Universidade Católica do Paraná, Brazil
António Barros	University of Porto, Portugal
Andreas Behrend	Technical University of Cologne, Germany
Karim Benouaret	Université Claude Bernard Lyon 1, France
Gloria Bordogna	CNR - National Research Council, Italy
Omar Boussaid	University of Lyon 2, France
Nieves R. Brisaboa	Universidade de A Coruña, Spain
Michael Brodie	Harvard Univeristy, USA
Francesco Buccafurri	University of Reggio Calabria, Italy
Alina Campan	Northern Kentucky University, USA
Arif Canakoglu	Policlinico di Milano, Italy
Antonio Corral	University of Almería, Spain
Gianni Costa	ICAR-CNR, Italy
Bruno Defude	Télécom SudParis, France
Paolino Di Felice	University of L'Aquila, Italy
Rebeca Díaz Redondo	University of Vigo, Spain
Fabien Duchateau	Université Claude Bernard Lyon 1/LIRIS, France
John Easton	University of Birmingham, UK
Georgios Evangelidis	University of Macedonia, Greece
Luiz Paulo Fávero	University of São Paulo, Brazil

João Ferreira	ISCTE, Portugal
Pedro Ferreira	University of Porto, Portugal
Valerio Frascolla	Intel, Germany
Pedro Furtado	University of Coimbra, Portugal
Vasco Furtado	University of Fortaleza, Brazil
Jérôme Gensel	Université Grenoble Alpes, France
Ross Gore	Old Dominion University, USA
Fabio Grandi	University of Bologna, Italy
Giulio Di Gravio	Sapienza University of Rome, Italy
Jukka Heikkonen	University of Turku, Finland
Victoria Hodge	University of York, UK
Jang-Eui Hong	Chungbuk National University, Korea, Republic of
Mark Hwang	Central Michigan University, USA
Sergio Ilarri	University of Zaragoza, Spain
Antonio Irpino	University of Campania L. Vanvitelli, Italy
Nikos Karacapilidis	University of Patras, Greece
David Kensche	SAP, Germany
Ravindra Khattree	Oakland University, USA
Attila Kiss	Eötvös Loránd University, Hungary
Mieczyslaw Kokar	Northeastern University, USA
Kostas Kolomvatsos	National and Kapodistrian University of Athens, Greece
Koji Koyamada	Kyoto University, Japan
Vladimir Kurbalija	University of Novi Sad, Serbia
Jose Labra Gayo	Universidad de Oviedo, Spain
Arijit Laha	Infosys, India
Thierry Lecroq	University of Rouen Normandy, France
Raimondas Lencevicius	Nuance Communications, USA
Maurizio Lenzerini	Università di Roma La Sapienza, Italy
Andrea Marrella	Sapienza University of Rome, Italy
Daniel Martinez-Avila	Universidad Carlos III de Madrid, Spain
Miguel Martínez-Prieto	University of Valladolid, Spain
Paulo Martins	Universidade de Trás-os-Montes e Alto Douro, Portugal
Arghir-Nicolae Moldovan	National College of Ireland, Ireland
Stefano Montanelli	Università degli Studi di Milano, Italy
Martin Necaský	Charles University in Prague, Czech Republic
Erich Neuhold	University of Vienna, Austria
Jens Nimis	University of Applied Sciences Karlsruhe, Germany
Riccardo Ortale	ICAR-CNR, Italy

Jan Paralic	Technická univerzita v Košiciach, Slovak Republic
Antonio Perianes-Rodriguez	Universidad Carlos III de Madrid, Spain
Jan Platos	VSB - Technical University of Ostrava, Czech Republic
Franck Ravat	IRIT-Université Toulouse I, France
Peter Revesz	University of Nebraska-Lincoln, USA
Colette Rolland	Université Paris 1 Panthéon-Sorbonne, France
Mercedes Ruiz	University of Cádiz, Spain
Gunter Saake	Institute of Technical and Business Information Systems, Germany
Melike Sah	Near East University, Turkey
Raazesh Sainudiin	Uppsala Universitet, Sweden
C. Okan Şakar	Bahçeşehir University, Turkey
Alberto Sanchez	Universidad Rey Juan Carlos, Spain
Iulian Sandu Popa	University of Versailles Saint-Quentin-en-Yvelines & Inria Saclay, France
Ahmet Sayar	Kocaeli University, Turkey
Fabio Schreiber	Politecnico di Milano, Italy
Diego Seco	Universidade da Coruña, Spain
Hayri Sever	Çankaya University, Turkey
Fatimah Sidi	Universiti Putra Malaysia, Malaysia
Cristina Sisu	Brunel University London, UK
Spiros Skiadopoulos	University of the Peloponnese, Greece
Andrzej Skowron	University of Warsaw, Poland
Yannis Stavrakas	ATHENA Research Center, Greece
Alfred Stein	University of Twente, Netherlands
Armando Stellato	Tor Vergata University of Rome, Italy
Sergey Stupnikov	IPI RAN, Russian Federation
Zbigniew Suraj	University of Rzeszow, Poland
George Tambouratzis	Institute for Language and Speech Processing, Greece
Horia-Nicolai Teodorescu	Gheorghe Asachi Technical University of Iași, Romania
George E. Tsekouras	University of the Aegean, Greece
Jeffrey Ullman	Stanford University, USA
Jari Veijalainen	University of Jyväskylä, Finland
Marco Villani	University of Modena and Reggio Emilia, Italy
Gianluigi Viscusi	Linköping University, Sweden
Zeev Volkovich	ORT Braude College, Israel
Qingguo Wang	Meharry Medical College, USA

Leandro Wives	Universidade Federal do Rio Grande do Sul, Brazil
Shaomin Wu	University of Kent, UK
Shengkun Xie	Toronto Metropolitan University, Canada
Marcia Zeng	Kent State University, USA

Additional Reviewers

Saïd Abdeddaïm	LITIS, Université de Rouen Normandie, France
Zouhaier Brahmia	MIRACL Laboratory, Tunisia
Nikolaos Giarelis	University of Patras, Greece
Ilias Siachos	University of Patras, Greece
Mihalis Tsoukalos	University of the Peloponnese, Greece

Invited Speakers

Maurizio Lenzerini	Sapienza University of Rome, Italy
Luís Paulo Reis	University of Porto, Portugal

Contents

Balancing the Virtual World: Exploring the Efficacy of a Virtual Layer in Mitigating Cyber Sickness

Jonathan Harth[1](✉), Christian-Norbert Zimmer[2], and Michaela Zupanic[1]

[1] Universität Witten/Herdecke, Alfred-Herrhausen-Straße 50, 58455 Witten, Germany
jonathan.harth@uni-wh.de

[2] Mixed Reality and Visualization (MIREVI), University of Applied Sciences Düsseldorf, 40233 Düsseldorf, Germany

Abstract. Cyber Sickness (CS), a discomfort experienced by 30–80% of virtual environment users, poses a significant barrier to the widespread adoption of Virtual Reality (VR) technology. This study evaluates the potential of software modifications to mitigate CS symptoms. Specifically, it examines a software layer inspired by the "Seetroën Glasses" prototype. 80 students were subjected to VR roller coaster simulations to test the efficacy of this layer. The findings indicate that the software layer not only improved user performance but also prolonged the duration before participants opted to discontinue the VR experience by approximately two rounds. Although the layer does not grant complete immunity to CS, it notably postpones the onset of symptoms. The research shows that the virtual test environment is suitable for investigating CS and suggests that the software layer prototype holds promise for diminishing the adverse effects associated with CS.

Keywords: Cyber sickness · Virtual reality · Mixed methods · User study · Software layer · Seetroën glasses · Evaluation study

1 Introduction

Immersive interfaces such as simulators or virtual environments enable impressive interactions and compelling user experiences, but they carry the risk of severely compromising subjective well-being [1, 2]. The problem is known for a long time and by many names, e.g., Simulator Sickness (SiS) or Cyber Sickness (CS) [3].

Modern simulation systems such as the use of Virtual Reality (VR) can also trigger the above-mentioned complaints. Cyber Sickness is a condition that can occur during or after exposure to virtual environments. Symptoms include headaches, eye strain, nausea or even vomiting in extreme cases [4]. It is estimated that some degree of Cyber Sickness occurs in about 30% to 80% of all users [5]. Although the phenomenon has been known for a long time under the names of Simulator Sickness or Motion Sickness and is attributed to conflicts between the visual and vestibular systems [6], there is still no generally recognized solution for the prevention of the disease that is suitable for

O. Gusikhin et al. (Eds.): DATA 2023, CCIS 2105, pp. 1–24, 2024.
https://doi.org/10.1007/978-3-031-68919-2_1

everyday use [7, 8]. Thus, as long as these negative effects in the use of Virtual Reality cannot be reliably avoided, they will continue to be a major obstacle to the growth of the VR medium.

The quest to overcome this obstacle has led researchers and developers to explore various avenues, from hardware modifications to software solutions [9]. This paper addresses the latter by presenting an evaluation of a novel software solution aimed at mitigating the symptoms of Cyber Sickness.

The approach taken by this study is rooted in the principles of user-centered design, recognizing that the success of any intervention is contingent upon its acceptability and effectiveness from the perspective of the end-user [10]. Drawing inspiration from the innovative "Seetroën Glasses", which have made strides in addressing motion sickness in automotive settings, our research team has developed a software layer prototype that integrates similar principles into the VR environment.

In this version of our article, which was originally published in the conference proceedings of DATA 2022 [11], we have made significant additions and elaborations. Firstly, we have considerably expanded the chapter on related work and therefore embed the study much more strongly in the context of the current state of research. In addition, we have integrated the results of a pilot study at the SRH University of Applied Sciences in Heidelberg. These new data allow us to supplement and expand our previous findings, which contributes to a more comprehensive understanding of the phenomenon. Furthermore, we have included more pictures and information about the test environment and conducted more thorough analysis in this revised version. This includes extended comparisons that allow the results to be interpreted and discussed in a broader context, particularly in relation to the ataxia tests. Overall, this revised version of the paper provides a broader and deeper insight into our research and represents a significant contribution to our understanding of potential effects to mitigate Cyber Sickness.

2 Related Works

2.1 From Motion Sickness to Cyber Sickness

The study of Cyber Sickness, an adverse response to the immersive environments of Virtual Reality, continues to evolve as research advances our understanding of its symptoms and contributing factors. Extensive research on CS endeavors to deepen the comprehension of its symptoms and the elements contributing to its onset [5, 12]. The "sensory mismatch" theory posits that the primary genesis of CS is a discordance between visual stimuli and the corresponding vestibular or proprioceptive responses [13]. Today, a wide variety of elements are seen as impacting factors in inducing CS. These elements can be categorized into 1) hardware-dependent facets, such as issues with tracking, input/output delays, display refresh rates, or screen tearing, 2) content-oriented facets like the visual depiction of motion direction or the existence of linear or angular accelerations, and 3) human-dependent facets, such as age, gender, physicality, prior experience with virtual contents and more.

The phenomenon identified as CS, alternatively termed VR sickness, was first termed 1992 by McCauley and Sharkey [14]. Chang et al. [1] expounded that when visual stimuli serve as the predominant sensory input leading to motion sickness, this specific symptom is termed Visually Induced Motion Sickness (VIMS) [15]. VIMS has been referred to by various names depending on the context, including Gaming Sickness, Simulator Sickness, Cinerama Sickness, and VR-Sickness. Intriguingly, instances of Cinerama Sickness were reported as early as the 1900s [16].

The exploration of Motion Sickness (MoS) traces back to mid-twentieth-century research. The initial studies addressing this ailment, triggered by motion stimuli, are dated to 1945 in the Scopus database [17]. MoS generally manifests in individuals aboard moving vehicles or apparatus [18]. It is a disorder induced by motion stimuli in respective situations. In 1970 Money [19] summarized the accompanying symptoms as pallor, cold sweat, nausea, and vomiting. Motion perception can occur in a physical or a virtual space through vestibular movements or visual illusion. Visually Induced Motion Sickness (VIMS) is a term employed when this is triggered by a virtual environment [20].

With the utilization of simulators in diverse research fields, such as aerospace, a variant of MoS, named Simulator Sickness (SiS), was identified, which is triggered using a moving simulation apparatus [21]. Discrepancies between the displayed movement in the simulation and the movements generated by the apparatus can induce symptoms akin to regular MoS, including headaches, sweating, dry mouth, drowsiness, disorientation, dizziness, nausea, and vomiting [21]. Davis et al. and Milleville-Pennel et al. expanded this list to include apathy and eye strain [18, 22].

The advent of digitization introduced another form of Motion Sickness: Cyber Sickness (CS). McCauley et al. postulated that Cyber Sickness pertains to the illusion of body movement in a virtual environment and the consequent MoS symptoms [14]. This notion is supported by Kennedy et al., suggesting that a sense of movement can occur even without a physically perceived impulse [23]. Viewing a visual representation alone can induce symptoms, leading Kennedy et al. to define this ailment as visually induced Motion Sickness. Another perspective posits that MoS-like symptoms occurring while using Virtual Reality systems with Head-Mounted-Displays (HMD) or 3D screens can be categorized as Cyber Sickness [5].

All these definitions share the common prerequisite of a visual stimulus. If an individual closes their eyes in such situations, unlike conventional Motion Sickness, no manifestation of the ailment will occur [20]. The symptoms induced by Cyber Sickness largely resemble those of Simulator Sickness and classic Motion Sickness: disorientation, dizziness, eye strain, headaches, sweating, fatigue, and nausea [24]. Additionally, oculomotor disturbances, i.e., disruptions in eye movement, and Ataxia, a disorder of movement coordination, are listed.

Regarding the causes of these various motion sicknesses, diverse theories have been postulated. The most frequently proposed explanatory approach is found in the "Sensorimotor Conflict Theory" (SCT). This theory, first introduced by Reason and Brand, posits that the emerging symptoms are due to a discrepancy in motion perception [16]. Both the visual sense, represented by the eye, and the vestibular system contribute to detecting movement. The latter comprises the three semicircular canals and the two otolith organs, Sacculus and Utriculus. The otolith organs detect linear acceleration, gravitation, and head tilt, while the semicircular canals perceive angular acceleration [25]. When these two systems relay differing stimuli, the manifestation of MoS diseases occurs [4].

According to Chang et al., the inception of a VR system involves a complex amalgamation of hardware technologies and content rendering, suggesting the plausibility of numerous components contributing to user discomfort [1]. It has been further discerned that individual variations can influence the degree of VR sickness experienced [26, 27]. Despite users being exposed to identical VR content through the same device, the level of discomfort experienced is contingent on individual characteristics. Subsequent studies have endeavored to identify the human factors serving as prominent indices for predicting the severity of user discomfort, with factors such as motion sickness history and prior VR experience being extensively scrutinized [10, 28, 29]. There are also various contextual factors, such as room temperature, humidity, and ambient noise [24]. Accordingly, Chang et al. identified three primary causes of VR sickness, namely, hardware, content, and human factors [1]. The effects of device-related features like display types and modes were extensively investigated under hardware factors. Content factors encompassed various aspects related to the VR scene, including optical flow, graphic realism, rendering reference frames, and task-related features, while human factors incorporated several demographic features such as age, gender, and motion sickness history.

Contrary to LaViola's perspective, Rebenitsch and Owen have posited that technological advancements do not necessarily correlate with a decrease in Cyber Sickness [4]. They put forth a compilation of alternative factors contributing to the phenomenon. Moreover, a comprehensive meta-analysis by Saredakis et al. delved into the relationship between virtual content and specific Cyber Sickness side effects [7]. Their conclusion highlighted that the nature of virtual content itself could be a pivotal factor inducing Cyber Sickness. This infers that the emergence of Cyber Sickness isn't exclusively contingent on the technological quality (such as field of view and graphic fidelity), but also on the way content is presented to the user.

Similarly, MacArthur et al. recently underscored the importance of incorporating and documenting women's experiences to enhance understanding of gender-specific effects [30]. Concurrently, the study by Howard and van Zandt delineated a variety of human factors significantly related to Cyber Sickness, including susceptibility to motion sickness, gender, and real-world VR experience [31]. However, they documented inconsistent results across diverse individual differences. Furthermore, a seminal synthesis by Stanney et al. encapsulated critical studies from both past and present, proposing a forward-looking agenda aimed at mitigating Cyber Sickness [32].

Furthermore, perceptual processing exhibits variation across individuals. Field dependence characterizes the extent to which an individual's perception of their bodily position in space is swayed by contextual information [33]. Those with elevated levels of field dependence are more reliant on external cues, such as reference frames or background objects, whereas field-independent individuals predominantly depend on their internal vestibular and proprioceptive cues for perceiving body position. Field independence is more prevalent among males and correlates with a reduced propensity for motion sickness [34]. Field dependence can influence the perception of vection, the illusion of self-motion in virtual space, particularly when a visual frame is absent, and the earth's vertical is not distinctly represented [35].

In the investigation of Motion Sickness and related ailments, several factors potentially influencing individual susceptibility have been identified and examined across various studies. These encompass personal and individual factors, such as the gender of individuals experiencing the symptoms. While Biocca posits a gender difference to the detriment of women [36], Mittelstaedt asserts that although most studies identify a difference, it could be attributed to biased response behavior [37].

The concept of field dependence was initially introduced by Witkin, in collaboration with Asch, as "Field-dependence-independence" during their four-part study on individual spatial orientation [38, 39]. Witkin described field-dependent perception as being strongly influenced by the overall composition of the observer's environment, whereas field-independent individuals perceive separately from it [40]. In essence, field dependence can be defined as a preference to orient oneself using external visual cues in the environment, while field independence denotes a tendency to rely on internal vestibular cues [41]. Visual perception encompasses everything captured through the visual apparatus, while the vestibular system, located in the inner ear, is responsible for spatial orientation of the head [4]. Over time, various testing methods have been developed to measure a person's relative dependence on the field. This work introduces the methods applied in the studies under consideration.

In the Rod-Frame Test (RFT), the subject is situated in a completely darkened room, facing an illuminated rod within an illuminated square frame. The task is to align the rod with the vertical axis of gravity. There are variations where the head or even the entire body can be tilted, achieved, for example, by a tiltable chair. The illuminated frame can also be inclined [38, 39]. This setup creates a discrepancy between what the body sees and feels. Field-independent individuals find it easier to align the rod correctly, as they rely on their vestibular cues and perceive the gravitational axis better. Conversely, field-dependent individuals orient themselves to the visual cues provided by the frame, leading to more alignment errors.

Witkin hypothesized that field-independent individuals can perceive their environment more analytically and separately from environmental factors. In a study, he found that individuals struggling to extract simple figures from complex ones also had difficulty separating body inclination in the RFT from the actual gravitational vertical [42]. Therefore, he developed the Embedded Figures Test (EFT) based on the figures of Gottschaldt, who researched the influence of prior experience on perception [43]. In this test, the task is to identify a simple figure embedded within various complex figures. The test comprises 24 complex figures, and to compare participants' results, the time taken to solve the tasks is measured [44]. Field-independent individuals find it easier to extract simple figures from the complex environment and accordingly, take less time to complete the tasks compared to field-dependent individuals. A recent study for NATO also concludes that further research on field dependence would be necessary if economically justifiable [45].

In summary, the literature reflects a rich tapestry of research efforts seeking to unravel the complexities of Cyber Sickness. Contrary to earlier studies, recent publications suggest that Motion Sickness and Cyber Sickness can be regarded as the same clinical condition, with Cyber Sickness being a subcategory of Motion Sickness [13]. From the foundational theories of sensory conflict to the nuanced understanding of individual differences, the works collectively suggest a convergence of factors that elicit this condition. The exploration of Motion Sickness (MoS), Simulator Sickness (SiS), and

Cyber Sickness (CS) elucidates a shared underpinning in visual stimulus and sensory discordance [46]. Technology may advance, yet the presentation of the virtual content itself, gender-specific experiences, and the array of human factors weave a complex web that thwarts one-size-fits-all solutions [5, 10, 30].

2.2 Seetroën Glasses

The challenge of motion sickness in dynamic environments has led to innovative solutions aimed at mitigating its uncomfortable symptoms. An exemplar of such ingenuity is the Seetroën Glasses, introduced by Citroën, which offer a novel approach to combating motion sickness in various settings, from vehicular to maritime travel.

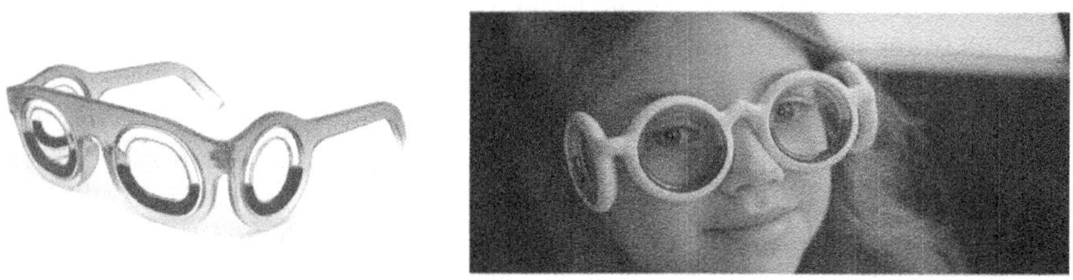

Fig. 1. Seetroën glasses by car manufacturer Citroën [47].

The glasses were developed in collaboration with Boarding Ring, a French start-up specializing in motion sickness solutions, and employ a patented technology that introduces a liquid-filled ring around the eyes, visible in the peripheral vision. This design creates a dynamic, artificial horizon within the rings, operating along both the frontal (left/right) and the sagittal (front/back) planes. The Seetroën Glasses work by realigning the sensory information between the eyes and the inner ear, where the conflict that leads to motion sickness arises [48]. Citroën claims a remarkable 95% efficacy rate, suggesting that the glasses can significantly alleviate symptoms within 10 to 12 min of use, and once the equilibrium is restored, they can be removed [49] (Fig. 2).

Fig. 2. Boarding Ring Glasses [50].

Despite the claimed success rate, independent research verifying the efficacy of the Seetroën Glasses is scant. Nonetheless, anecdotal evidence and subjective tests suggest that the glasses are indeed effective, particularly in urban environments. Users have reported relief from motion sickness without the need to maintain a fixed gaze down streets or horizons, contrasting with the discomfort experienced by those not using the glasses.

Remarkably, accordingly to Citroën over 40.000 pairs of the Seetroën Glasses have been distributed, indicating a substantial user base, and suggesting widespread consumer satisfaction [47]. The glasses are designed to be versatile and lens-free, making them suitable for both adults and children over the age of ten—the developmental stage when the inner ear is typically fully grown. This design also allows for easy sharing among family members and travel companions. Citroën has also released a limited edition, the "Seetroën S19", to celebrate the brand's centenary, producing only 1,919 units.

In the context of this study, we seek to examine how a software-based adaptation of this physical concept can be employed to reduce discomfort in Virtual Reality applications. The aim is to replicate the success of the Seetroën Glasses within the digital domain by introducing a virtual layer over the VR imagery, thereby delaying the onset of Cyber Sickness.

2.3 Measuring Cyber Sickness

To assess Cyber Sickness symptoms, a variety of measurement tools have been devised over time. The most popular and prominent tools are Motion Sickness Questionnaire, Simulator Sickness Questionnaire (SSQ), Cyber Sickness Questionnaire, Virtual Reality Sickness Questionnaire, Visually Induced Motion Sickness Susceptibility Questionnaire, and the Fast Motion Sickness Questionnaire. However, there is no generally acknowledged questionnaire for measuring sickness in virtual environment. Querying individuals about their experiences with Motion Sickness is the most utilized method [51]. In 1968 Reason initiated the first attempt to develop such a questionnaire with the Motion Sickness Questionnaire (MSQ) [52], which inquires participants about their past experiences with the ailment, divided into Section A) age range up to twelve and B) age group above. It includes queries about nausea or vomiting during the use of various vehicles (cars, buses, trains, airplanes, ships) and amusement rides (large swings, carousel). Section B was later modified by Golding to cover the frequencies of experiences during the last ten years [53].

The Simulator Sickness Questionnaire (SSQ) by Kennedy et al. has been the most frequently used tool to determine motion sickness symptoms in virtual environments [54]. It consists of 16 items, which encompass negative effects of using a simulator on three symptom-specific subscales (gastrointestinal, visuomotor, and vestibular symptoms). All these measurements are subjective as they rely on participants reporting their own experiences.

Sevinc and Berkman asses various new measurement tools, which inquire more explicitly about Cyber Sickness symptoms [55]. These are the Cyber Sickness Questionnaire [56] and the Virtual Reality Sickness Questionnaire [57]. Another promising approach is presented by the Visually Induced Motion Sickness Susceptibility Questionnaire [58].

The Fast Motion Scale (FMS) was first introduced in the work Validating an Efficient Method to Quantify Motion Sickness [15]. In this work, the authors sought another way to measure motion sickness. They tested a new method called Fast Motion Sickness Scale with 126 subjects to be able to measure motion sickness during the experience. They compared the results with the then much used Simulator Sickness Questionnaire. In the case of the FMS, participants are asked to verbally communicate their level of Cyber Sickness, for example, on a scale of 0–20 [15] or 1–10[59]. This allows for the construction of a continuous table or graph representing the subjective well-being of the participant. Consequently, segments of the application that induce an elevated level of Cyber Sickness can be identified [60]. The FMS was initially employed in 2011 to ensure data collection during immersion. Since then, FMS has often been found as an additional means alongside questionnaires in recent studies [55, 60] and has proven to be a valid tool.

The Cyber Sickness Questionnaire (CSQ) primarily evolved from the Simulator Sickness Questionnaire (SSQ) after the latter was deemed unsuitable for the VR context in several studies. As a result, the CSQ has been recurrently employed in contemporary research [55, 56] and is a frequently utilized tool for data collection. This utilization allows the results obtained from two-sample t-tests and Pearson correlations to be compared and evaluated with these studies. A potential source of error in the CSQ lies in the omitted symptoms that are still present in the SSQ. As stated, the CSQ consists of 9 out of 16 symptoms from the SSQ. These were chosen as clear indicators of Cyber Sickness. However, during the study, participants reported symptoms not covered by the CSQ, such as sweating on hands and forehead, and increased salivation, which are, however, included in the SSQ.

3 Methods

3.1 Participants

In the study, eighty students participated, comprising 49 females and 31 males, with an average age of 23.51 ± 5.89 years (ranging from 18 to 45 years). They were randomly allocated to one of two groups: control or layer. The degree of field dependence was measured using the three visual perception subtests 6–8 of the LPS-2. The individual raw scores (on average subtest 6: 23.94 ± 6.91; subtest 7: 26.64 ± 4.59; subtest 8: 23.94 $\pm$ 6.91) being summed up and converted into norm scores to an intelligence test scale with 100 ± 15 (Mean $\pm$ SD) [61]. The average norm score for performance in the LPS-2 Visual Perception was 101.50 ± 14.73 (with a range of 74 to 139). The distribution is slightly right skewed (skewness = .217) and flattened (kurtosis = −.585).

Participants with a score below the median of 100 were categorized as field-dependent, while those scoring above 100 were deemed field-independent, following a median dichotomization of performance in the LPS-2 Visual Perception. Factoring in perceptual style and viewing conditions, participant results were structured in a quasi-experimental design, since the relevant influencing variable perceptual style is not directly controlled, but a subsequent control is carried out via statistical procedure [62]. Both female and male participants were uniformly distributed across the four groups (Fig. 3 and Table 1).

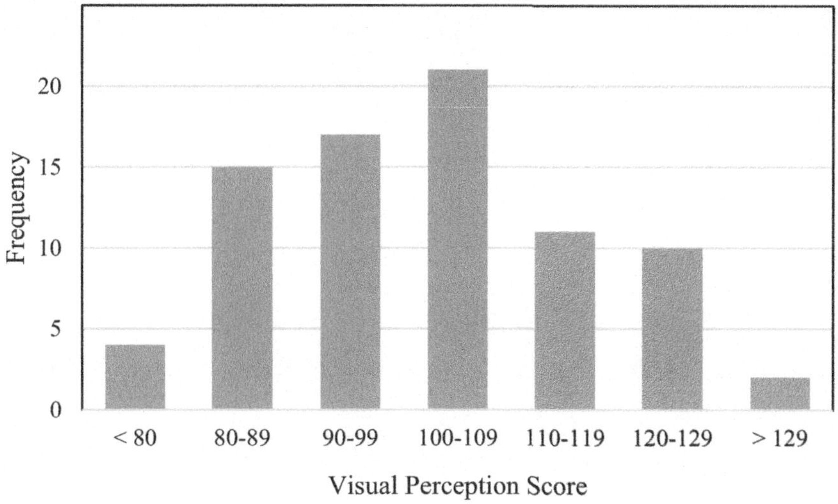

Fig. 3. Distribution of visual perception score (N = 80).

Table 1. Quasi experimental design with perception style and viewing conditions.

	Control (N = 40)	Layer (N = 40)
Field dependence low	N = 21 (10 f., 11 m.)	N = 20 (11 f., 9 m.)
Field dependence high	N = 19 (13 f., 6 m.)	N = 20 (15 f., 5 m.)

Participants' prior experience with motion stimulation was also explored. Approximately a third of the participants (N = 25, 31.3%) reported having ridden a roller coaster never or only once, 25% (N = 20) up to ten times, and the majority (N = 35, 43.8%) more than ten times. A weakly significant correlation with age became apparent. Older participants have less experience with the roller coaster than younger ones (Spearman Correlation: $r = -.248$, $p < .027$).

Regarding frequency of playing computer games, half of the participants (N = 42, 52.5%) reported never playing, followed by those playing at least once a month (N = 19, 23.8%), at least once a week (N = 15, 18.8%), and daily (N = 4, 5%). There was a clear gender effect (Chi Square = 17.34, $p < .001$), as 69.4% (N = 34) women reported never playing computer games, while only 25.8% (N = 8) men did. A mere 4 participants (5%) engage with VR at least once a month (Fig. 4).

The participants' subjective sensitivity to motion, rated on a scale from 0–10, yielded an average score of 4.18 ± 2.34 (ranging from 0 to 9) for the sample. Consequently, participants tended to perceive themselves as relatively insensitive to motion, but women (4.41 ± 2.44) report higher levels of subjective sensitivity to motion than men ($3.81 \pm$

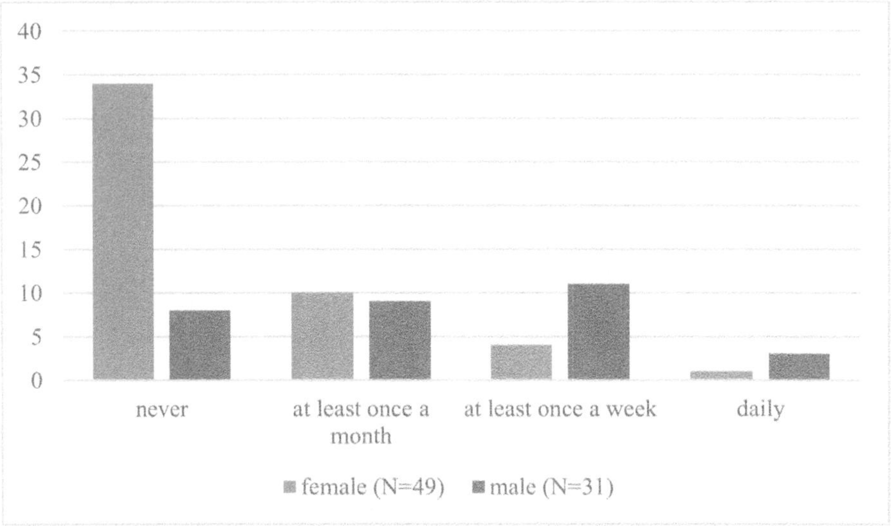

Fig. 4. Ratings of playing computer games depending on gender (N = 80; 49 female, 31 male).

2.17) (Mann-Whitney-U = 211, p = .018). Participants with higher sensitivity used a roller coaster less often than insensitive participants (Spearman Correlation: r = −.303, p < .006).

3.2 The Virtual Layer

For this study, both the application and the layer were meticulously developed from the ground up utilizing the Unity Engine. The conception of the virtual layer drew inspiration from the principle underpinning Citroën's "Seetroën glasses" (Fig. 1). Fundamentally, the layer was designed with versatility in mind, aiming for easy integration into any VR application. To assess the efficacy of the layer, an application with a propensity to induce CS was developed.

In collaboration with the Department of Mixed Reality and Visualization (MIREVI) of the University of Applied Sciences Düsseldorf, a prototype was crafted, offering users peripheral fixed points within the head-mounted display (HMD) reminiscent of the "Seetroën Glasses". Realized through the Unity engine, the application embodies the following features:

- A roller coaster ride, enduring 70 s per lap,
- Enablement/Disablement of continuous rides,
- Activation/Deactivation of the VR layer,
- Adjustable offset for the layer (0–1),
- Variability in the virtual distance of the layer from the user's POV (2 m, 1 m, 50 cm, 30 cm, 20 cm, 10 cm).

The study's test environment immerses participants in a virtual roller coaster ride, incorporating a looping and multiple steep curves and inclines, meticulously designed to induce CS. Fast-paced travel through tight curves and overhead rides are deemed conducive to eliciting the symptoms under scrutiny [63]. The hardware setup for the study incorporated a VR-capable computer (AMD Ryzen 5600, GeForce RTX 3080, 32 GB RAM) and a compatible VR Headset (Valve Index) (Figs. 5 and 6).

Fig. 5. Point of View (POV) of a user riding the Roller Coaster with activated Layer [11].

Fig. 6. Overview of the virtual Roller Coaster with one looping an several narrow slopes.

Through the implementation of a virtual visual layer, modeled after the glasses, the study aims to mitigate users' visual-vestibular conflicts. The pivotal research question centers around the extent to which the principles of the anti-motion sickness "Seetroën Glasses" can be adapted for VR headsets, and subsequently, the degree to which the layer serves as a remedy for CS symptoms.

3.3 Procedures

Employing a mixed methods approach, we executed a quasi-experimental study integrating questionnaire surveys, static ataxia tests, and subsequent qualitative interviews for comprehensive evaluation. The questionnaires were instrumental in establishing an objectifiable relationship between the levels of CS and the different prototypes, while the interviews primarily facilitated a subjective articulation and assessment of the experiences from the participants' perspective. Eligibility criteria mandated a minimum age of 18 years and excluded individuals with severe health conditions (such as epilepsy, significant cardiovascular diseases, hearing organ disorders), those with pacemakers or other implanted medical devices, and those with acute or chronic mental illnesses.

The study secured approval from the Ethics Committee of Witten/Herdecke University (S-142/2021), and subjects were informed of their right to withdraw participation at any point, as highlighted in the consent form. Moreover, the study adhered to the Code of Conduct on the Ethical Use of Virtual Reality in Research [64].

The Cyber Sickness Questionnaire (CSQ), developed drawing upon the Simulator Sickness Questionnaire and the Virtual Reality Sickness Questionnaire, served as a quantifiable survey instrument [65]. It comprised 24 items in total, including six demographic items (age, gender, prior VR experience) and nine items assessing symptoms of CS such as nausea, headaches, and dizziness, answered on a 4-point Likert scale (1 = none, 2 = mild, 3 = moderate, 4 = severe). The CSQ was administered to participants both before and after the roller coaster ride to facilitate a pre/post comparison.

Static ataxia tests, structured in a pre/post design, aimed to measure potential equilibrium disorders in participants induced by the roller coaster experience [66]. The tests required maintaining a specific body posture for 30 s, initially with eyes open and subsequently with eyes closed. These two modes of ataxia tests were performed in two runs each, first by standing on preferred leg with eyes closed (SOPLEC resp. SOPLEO) and then by standing on the nonpreferred leg with eyes closed/open (SONPLEC resp. SONPLEO). Each Performance was recorded before and after the ride.

The perceptual style of field dependence was assessed using three subtests from the performance test system "Leistungsprüfsystem 2" (Performance testing system, LPS-2) for visual perception: mental rotation, area count, and line pattern [61]. The theoretical foundation of the LPS-2 is in the context of Carroll's three-level model of intelligence [67]. LPS-2 subtests tests are classified according to the postulated G-factor of general intelligence: crystalline intelligence (GC), fluid intelligence (GF), visual perception (GV) and cognitive speed (GS). The selected three subtests for visual perception subtest 6: mental rotation, subtest 7: area count and subtest 8: embedded figures, which summed up represent the factor visual perception (GV). The internal consistency of these subtests ranged from good to very good, with $\alpha = .86$, $\alpha = .90$, and $\alpha = .93$. An example for an instruction for the embedded figures test (LPS-2 subtest 8) is shown in Fig. 7.

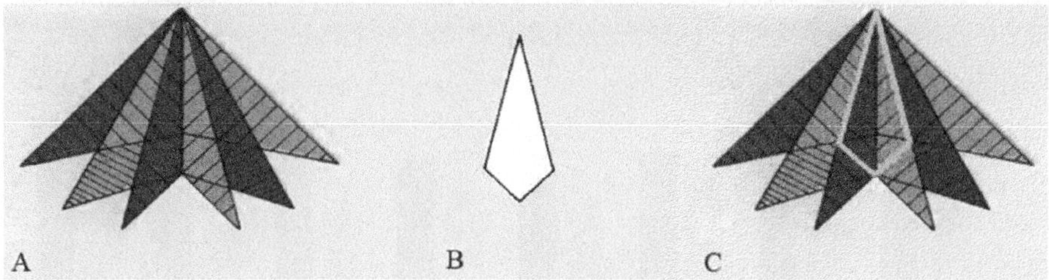

Fig. 7. An example of an item taken from the Embedded Figure Test (EFT). (A) The geometric colored complex figure. (B) The simple black and white simple shape. (C) The simple shape within the complex figure [68].

Furthermore, the Fast Motion Scale (FMS) was implemented during participants' immersion. This entailed acquiring a self-reported, subjective measure of CS, continually solicited by the study leader during immersion [60]. In the FMS, participants verbally communicated their level of CS on a predetermined scale, facilitating the generation of a running chart reflecting the subject's subjective well-being. This enabled the identification of specific segments of an application or process inducing heightened levels of CS. In our study, participants' well-being was recorded after each lap on a 0–10 scale. The total duration for each participant in the study approximated one hour.

4 Results

The outcomes of the study encompass both the foundational technical development and programming of the virtual layer in virtual reality, inspired by Seetroën glasses, as well as the testing and evaluation of the prototype through a randomized control trial. The prototype's successful creation was based on the "Seetroën Glasses" where, in VR, a layer could be activated over the visible image via the options menu. Consequently, test subjects were segregated into two groups: the control group experienced the roller coaster without a layer, while the experimental group had the tested layer in their field of vision.

4.1 Cyber Sickness Questionnaire

The theoretical examination of the questionnaire was conducted through item analyses, wherein the mean value, standard deviation, and discriminatory power of all items were calculated. As per Lienert/Raatz, the coefficient of selectivity denotes the correlation between the item response and the sum value of the scale [69], serving as an indicative value for the extent to which the differentiation of the individuals by the item aligns with that by the scale. Post-intervention, items 1 ("Do you suffer from nausea?"), 4 ("Do you suffer from dizziness when your eyes are closed?"), and 6 ("Does your head feel heavy or full?") exhibited the highest concurrence in the post-CSQ, with respective mean values of (M $\pm$ SD) 2.20 $\pm$ 0.82, 1.73 $\pm$ 0.69 and 1.73 $\pm$ 0.76). The discriminatory power of the items in the post-CSQ consistently exceeded the critical value of .20, with many situating within a good to very good range (r .60) (Fig. 8).

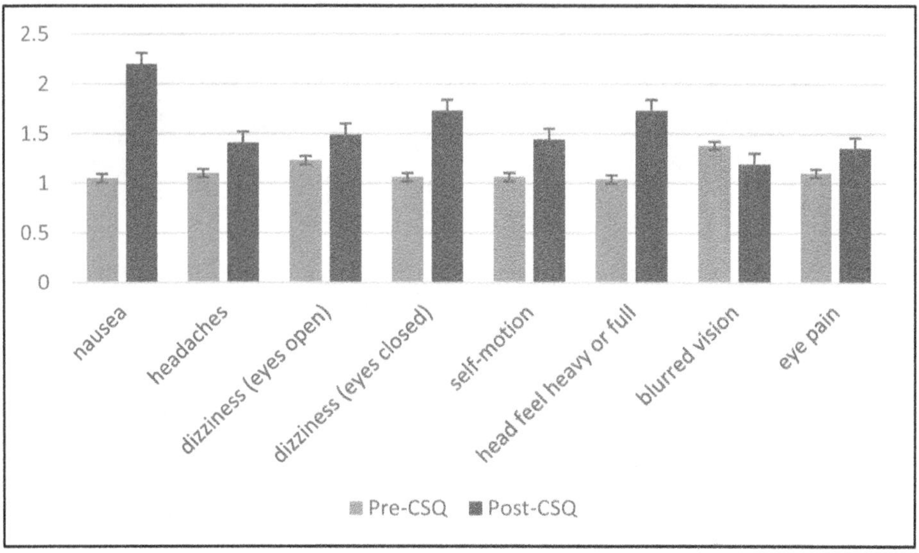

Fig. 8. Results for pre- and post-Cyber Sickness Questionnaire (N = 80).

These results are consistent to the results from the pilot study at SRH after 13 rounds of roller coaster. After the intervention, item 1 (Do you suffer from nausea?), 6 (Does your head feel heavy or full?) and 3 (Do you suffer from dizziness when your eyes are open?) have the highest agreement in the post-CSQ, with mean values of (M ± SD) 1.80 ± 0.91, 1.78 ± 0.83 and 1.65 ± 0.92 respectively. In the post-CSQ_sum scale (1.36 ± 0.51), the internal consistency is high (α > .800), whereas in the pre-CSQ-sum (1.09 ± 0.15), it is low (α >. 800).

To assess the CSQ scale, reliability was determined according to Cronbach's Alpha. This model of internal consistency is premised on the average inter-item correlation. Characteristic values of the CSQ scales include the scale mean (M), standard deviations (SD), and homogeneity coefficients (α). As anticipated, a higher concurrence was identified in the Post-CSQ_sum scale (1.57 ± 0.40) compared to the Pre-CSQsum (1.13 ± 0.22). Both scales manifested high internal consistency, deemed satisfactory to good (α > .700). These results are again similar to the results at SRH. In the Post-CSQ_sum scale (1.36 ± 0.51), internal consistency is high (α > .800), whereas in the Pre-CSQ-sum (1.09 ± 0.15) it is low (α > .800). Statistical evaluation of differences in the CSQ scales amongst the study groups (control vs. layer, field-independent vs. field-dependent, quasi-experimental design) unveiled no significant disparities.

4.2 Ataxia Tests

Analogous to the CSQ, the ataxia tests were executed before and after the roller coaster ride, each for a maximum duration of 30 s. The ataxia tests were performed in two modes and two runs each. The participants' performance in the two tests, SOPLEC (stand on preferred leg eyes closed) and SONPLEC (stand on non-preferred leg eyes closed), is catalogued in Table 3 (Table 2).

Table 2. Ataxia tests (SOPLEC, SONPLEC, pre/post).

Ataxia Tests	M	SD
Pre_SOPLEC_sum	13.99	8.97
Post_SOPLEC_sum	13.07	9.19
Pre_SONPLEC_sum	13.46	8.81
Post_SONPLEC_sum	13.31	8.12

Subsequent statistical examination disclosed a significant difference. Specifically, in the control group, the performance in the ataxia test SOPLEC was significantly less favorable post-roller coaster ride compared to the pre-test. Contrastingly, the group exposed to the layer demonstrated slightly enhanced performance ($F = 4.55$, $p = .036$) (Fig. 9).

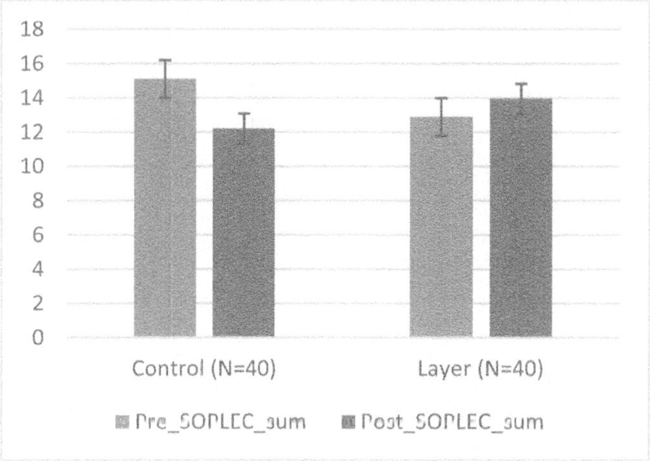

Fig. 9. Performance in pre/post SOPLEC ataxia test before/after the intervention (N = 80) [11].

4.3 Fast Motion Scale

The Fast Motion Scale (FMS) serves as a measure for assessing the subjective experience of CS during immersion, specifically throughout the 15 laps of the roller coaster ride in this study. Participants were given the option to conclude the immersion prematurely, resulting in only the initial two laps being universally completed. There was a significant correlation of negative proportionality between the number of laps and subjective sensitivity ($r = -.224$, $p = .009$). Participants who rated themselves as sensitive to motion sickness were more likely to abandon the roller coaster ride. The FMS exhibits a substantial spread of values, with a response scale ranging from 1 to 10. The average number of completed laps (minimum = 2 and maximum = 15) do not differ for the study groups control (8.53 ± 5.03) and layer (9.87 ± 4.48).

Table 3. Descriptive statistics for FMS items in 15 laps (N = 80).

Fast Motion Scale (FMS)	Control N	Control M	Layer N	Layer M
FMS_1	40	2.35	40	1.30
FMS_2	40	3.38	40	1.88
FMS_3	36	3.72	40	2.90
FMS_4	32	3.91	37	3.22
FMS_5	27	3.85	36	3.53
FMS_6	24	4.04	30	3.53
FMS_7	21	4.05	26	3.42
FMS_8	21	4.38	25	3.80
FMS_9	19	4.21	24	4.00
FMS_10	17	4.00	20	3.65
FMS_11	14	3.64	18	3.67
FMS_12	13	3.23	15	2.93
FMS_13	13	3.54	15	3.00
FMS_14	12	3.58	15	3.27
FMS_15	12	3.58	14	2.86

The swift decrement in the sample size during roller coaster ride is manifest in Fig. 10 for both groups. A direct comparison of exit times between the two groups shows that the exit of a participant with layer is delayed about two laps.

Conversely, the 1-survival function (Fig. 11) shows the progressive increase in Cyber Sickness over time, as delineated in existing literature. The initial third of dropouts occurs up to lap 5 (N = 26, 32.5%), the subsequent third nearly reaching the final lap (N = 28, 35%), and only the remaining third of participants (N = 26, 32.5%) successfully completing all 15 laps of the intervention. Clear distinctions in the sum of the CSQ post-intervention were observed between the three groups (Kruskal-Wallis-H = 21.72, $p < .001$).

These three distinct groups demonstrate significant variations not only in the number of laps completed but also in gender distribution, subjective sensitivity to motion, and the results of the Cyber Sickness Questionnaire (post-CSQ) following the intervention. The early dropout group contains significantly more females (N = 20) than males (N = 6) (Chi-Square = 6.73, $p = .035$). In the post-CSQ, this group exhibits the highest response to item 1 ("Do you suffer from nausea?") with a mean of 2.62 ± 0.64 (Kruskal-Wallis-H = 26.21, $p = .001$).

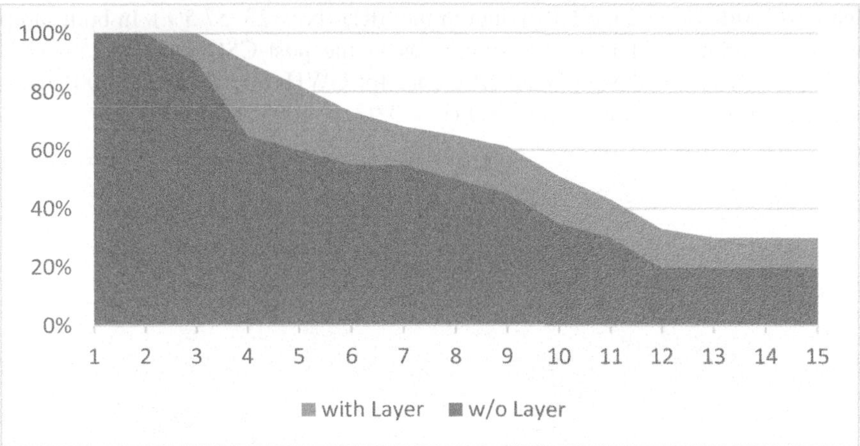

Fig. 10. Comparison of the cumulative exit points during 15 rounds of rollercoaster (with layer = blue; without layer = red; N = 80). (Color figure online)

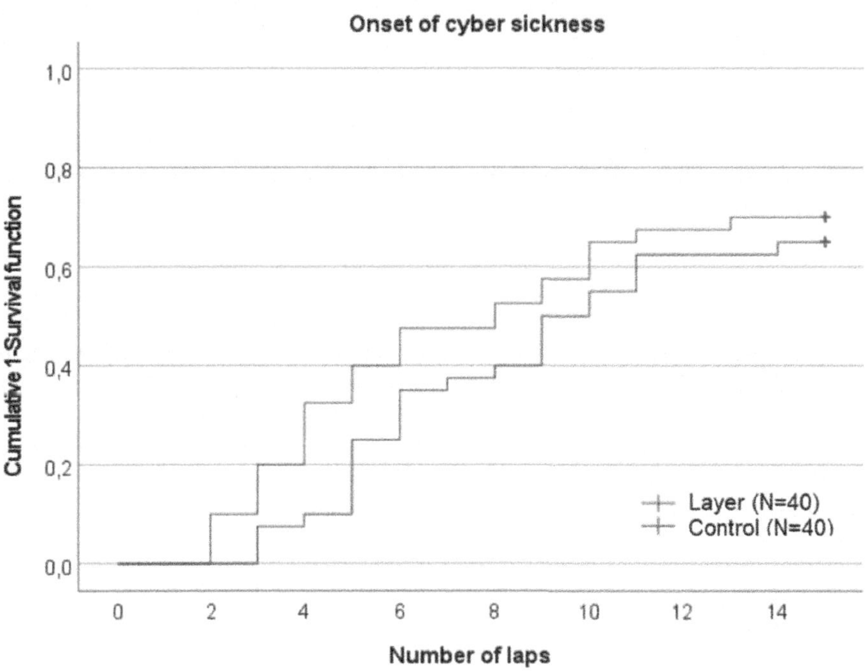

Fig. 11. 1-Survival function for the onset of Cyber Sickness in the roller coaster ride (N = 80).

For the direct comparison between the participants of the present study at Witten Herdecke University (UWH) and the pilot study at SRH University of Applied Sciences Heidelberg (SRH), the respective sample was divided into dropouts and completers. In the UWH sample, there were fewer participants in percentage terms who completed the rollercoaster ride (N = 26, 32.5%) than in the SRH (N = 23, 57.5%). In both samples, dropouts complained of more CS symptoms in the post-CSQ after 15 laps (UWH) resp. 13 laps (SRH), with significant difference for UWH ($F = 15.26$, $p < .001$) and just missed 5% probability of error for SRH ($F = 3.92$, $p = .055$) (Fig. 12).

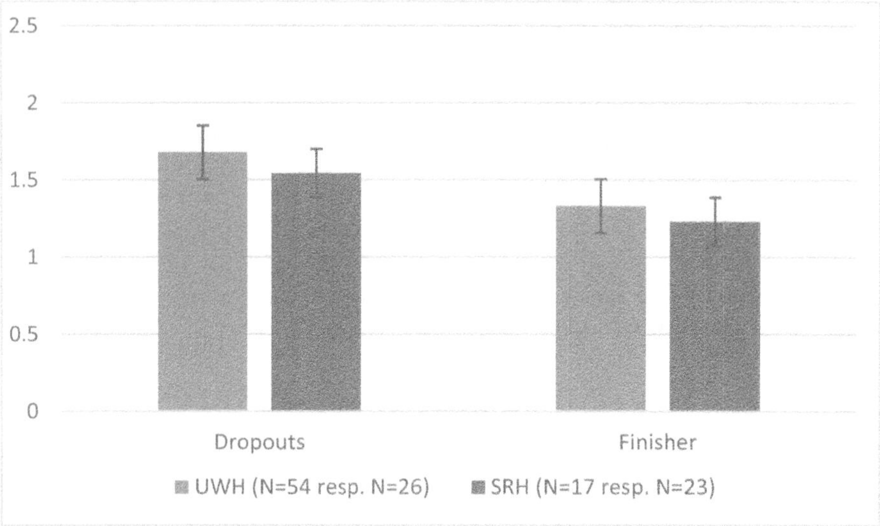

Fig. 12. Comparison of results in the Cyber Sickness Questionnaire (CSQ) after the intervention between pilot study and this study (N = 120).

To elucidate these effects statistically, all influencing factors - group (control vs. layer), field dependence, age, gender, and sensitivity to motion- are considered together as potential predictors of achieved FMS score in multiple regression analysis. The prediction of FMS score with gender and sensitivity as explanatory variables is significant ($F (2, 77) = 5.33$, $p < .007$). Gender clarified 5.7% of the variance ($\beta = 2.63$, $T = 2.41$, $p = .018$) and sensitivity to motion additional 4.2% ($\beta = -2.30$, $T = -2.14$, $p = .036$). The other variables were not included in the regression model. The significant correlation of positive proportionality between the sensitivity to motion and the FMS score ($r = .421$, $P < .001$) is a proof of content validity [70].

5 Conclusion

In general, a total of 80 participants (N = 80) were involved in the study. This number mirrors preceding studies [55] (N = 41) and therefore falls within a range of statistical relevance. The participants were divided into two groups: control and test group, the latter experiencing the application with the VR-layer. An alternative approach would have been to let the participants go through the prototype first without, then with the VR-layer. However, this approach was discarded as a potential habituation effect could not be ruled out. The structure of the study fundamentally followed that of previous studies [55, 56]. Hence, the decision was made to administer the CSQ not only once after immersion, as in most preceding studies, but also beforehand to exclude any pre-existing symptoms. Possible priming of the participants was accepted, given that motion sickness occurs even when the individual is aware that they are, for example, embarking on a sea or train journey.

Overall, the application proves to be an exceptionally apt testing environment for the onset of Cyber Sickness. Figure 11 lucidly depict the gradual intensification of CS over time—or throughout the duration of the roller coaster ride—as an increasing number of participants abandon the experience. Additionally, a distinct tripartite division emerges among the test subjects in our sample: a third terminate their participation by the fifth lap, another third endures until the tenth lap, and the final third exhibit no difficulty completing the entire session in VR. Closer scrutiny reveals that the sensitive group predominantly comprises female participants with limited experience in computer games. This observation aligns with existing research [7, 71], advocating for further research to better comprehend the prevalence of Cyber Sickness due to gender differences or other individual factors regarding immersive technology [1, 72]. Presently, the identified discrepancies are most coherently accounted for by a multifaceted interplay of diverse individual factors and uncontrolled experimental variables [63, 73].

The principal query of this study centered on whether the activation of the layer influences the incidence of CS. However, this technical assessment proved more challenging than anticipated. The interpretive value of our study's outcomes appears to be constrained by at least two limitations:

1. The transferability of our findings towards broader generalization is confined by the nature of the sample, characterized by its youthful demographic and heterogeneous background in digital games and virtual reality experiences. Future studies could benefit from an experimental design accommodating both gender differentiation and variance in computer game experience.
2. The methodological constraint stems from providing participants the liberty to exit at any juncture of the experiment, generating ethically commendable but less definitive data. It is plausible that participants terminated their involvement prematurely, prior to the manifestation of acute CS symptoms—especially true for the more sensitive subjects. Subsequent experimental series might refrain from offering a discretionary exit, awaiting instead the participants' explicit desire to withdraw. A more stringent approach, initially set aside primarily for ethical considerations, might have yielded higher lap counts, elevated FMS scores, and more pronounced disparities in the post-CSQ.

Despite these limitations, our data reveal a discernible trend regarding the layer's impact, aligning with existing literature on technical countermeasures for Cyber Sickness [74]. The direct comparison of exit times between experimental groups—with and without the layer—indicates that the layer potentially postpones participant withdrawal by approximately two laps (see Fig. 10). While the layer doesn't confer immunity, it does delay the onset of CS, suggesting its potential applicability in diverse virtual reality environments beyond the roller coaster scenarios. Further exploration is requisite to assess the layer's efficacy across varied settings and ascertain the consistency of its beneficial effects on mitigating Cyber Sickness symptoms.

It is also imperative to contemplate user acceptance of the layer. Although user satisfaction wasn't the focal point of our study, assessing whether users deem the layer a beneficial and acceptable addition to their virtual reality experiences is vital. Investigating optimal transparency levels for the layer and exploring adaptability to individual preferences and needs would yield invaluable insights for augmenting user experience.

The layer could be especially advantageous during the initial acclimatization phases of Virtual Reality applications, analogous to astronauts undergoing training to alleviate motion sickness. Employing the layer to mitigate the impact of Cyber Sickness during early exposures could facilitate users' gradual adaptation to the VR environment, diminishing dependency on the layer over time. This conceptualization frames the layer as a transitory aid for easing the immersion into VR, rather than a permanent feature.

In summation, our study signifies a commendable initial endeavor in investigating the virtual layer's potential as a mitigative tool for Cyber Sickness symptoms. While it doesn't offer absolute protection or immunity against the symptoms of Cyber Sickness, the virtual layer demonstrates a capacity to delay symptom onset, potentially providing meaningful benefits, especially during the introductory phases of virtual reality engagement. Cyber Sickness still poses a significant challenge to the widespread use of Virtual Reality, leading to different levels of discomfort and potentially breaking the immersive experience. That is why subsequent research endeavors are essential to refine the layer's application, gauge user acceptance, and evaluate its effectiveness across an extensive spectrum of virtual environments.

Acknowledgements. We would like to thank the Internal Research Grant of the Faculty of Health at Witten/Herdecke University for funding this study.

References

1. Chang, E., Kim, H.T., Yoo, B.: Virtual reality sickness: a review of causes and measurements. Int. J. Hum.-Comput. Interact. **36**(17), 1658–1682 (2020)
2. Mayor, J., Raya, L., Sanchez, A.: A comparative study of virtual reality methods of interaction and locomotion based on presence, cybersickness, and usability. IEEE Trans. Emerg. Top. Comput. **9**(3), 1542–1553 (2021)
3. Dennison, M.S.; Krum, D.M.: Unifying research to address motion sickness. In: 2019 IEEE Conference on Virtual Reality and 3D User Interfaces (VR), pp. 1858–1859 (2019)
4. LaViola, J.J.: A discussion of cybersickness in virtual environments. ACM SIGCHI Bull. **32**(1), 47–56 (2000)
5. Rebenitsch, L., Owen, C.: Review on cybersickness in applications and visual displays. Virtual Reality **20**, 101–125 (2016)
6. Stanney, K.M., Kennedy, R.S., Drexler, J.M.: Cybersickness is not simulator sickness. In: Proceedings of the Human Factors and Ergonomics Society Annual Meeting, vol. 41, no. 2, pp. 1138–1142 (1997)
7. Saredakis, D., Szpak, A., Birckhead, B., Keage, H.A.D., Rizzo, A.; Loetscher, T.: Factors associated with virtual reality sickness in head-mounted displays: a systematic review and meta-analysis. Front. Hum. Neurosci. **14**(96) (2020)
8. Yildirim, C.: Don't make me sick: investigating the incidence of cybersickness in commercial virtual reality headsets. Virtual Reality **24**, 231–239 (2020)

9. Oh, H., Son, W.: Cybersickness and its severity arising from virtual reality content: a comprehensive study. Sensors **22**, 1314 (2022)
10. Tian, N., Lopes, P., Boulic, R.: A review of cybersickness in head-mounted displays: raising attention to individual susceptibility. Virtual Reality **26**, 1409–1441 (2022)
11. Harth, J., Zimmer, C., Zupanic, M.: Look at the horizon: evaluation of a software solution against cyber sickness in virtual reality applications. In: Proceedings of the 12th International Conference on Data Science, Technology and Applications (DATA 2023), pp. 13–21 (2022)
12. Duzmanska, N.P., Strojny, P., Strojny, A.: Can simulator sickness be avoided? A review on temporal aspects of simulator sickness. Front. Psychol. **9**, 2132 (2018)
13. Gavgani, A.M., Walker, F.R., Hodgson, D.M., Nalivaiko, E.: A comparative study of cybersickness during exposure to virtual reality and "classic" motion sickness: are they different? J. Appl. Physiol. **125**(6), 1670–1680 (2018)
14. McCauley, M.E., Sharkey, T.J.: Cybersickness: perception of self-motion in virtual environments. Presence: Teleoper. Virtual Environ. **1**(3), 11–318 (1992)
15. Keshavarz, B., Hecht, H.: Validating an efficient method to quantify motion sickness. Hum. Factors **53**(4), 415–426 (2011)
16. Reason, J.T., Brand, J.J.: Motion Sickness. Academic Press, London (1975)
17. Alexander, S.., Cotzin, M., Hill Jr., C.J., Ricciuti, E.A., Wendt, G.R.: Wesleyan university studies of motion sickness: I. The effects of variation of time intervals between accelerations upon sickness rates. J. Psychol.: Interdisc. Appl. **19**(1), 49–62 (1945)
18. David, S., Nesbitt, K., Naivaiko, E.: A systematic review of cybersickness. In: Proceedings of the 2014 Conference on Interactive Entertainment (IE2014), pp. 1–9 (2014)
19. Money, K.E.: Motion sickness. Physiol. Rev. **50**(1), 1–39 (1970)
20. Koohestani, A., et al.: A Knowledge discovery in motion sickness: a comprehensive literature review. IEEE Access **7**, 85755–85770 (2019)
21. Brooks, J.O., et al.: Simulator sickness during driving simulation studies. Accid. Anal. Prevent. **42**(3), 788–796 (2010)
22. Milleville-Pennel, I., Charron, C.: Do mental workload and presence experienced when driving a real car predispose drivers to simulator sickness? An exploratory study. Accid. Anal. Prev. **74**, 192–202 (2015)
23. Kennedy, R.S., Drexler, J., Kennedy, R.C.: Research in visually induced motion sickness. Appl. Ergon. **41**(4), 494–503 (2010)
24. Bockelman, P., Lingum, D.: Factors of Cybersickness. In: Stephanidis, C. (eds) HCI 2017. CCIS, vol. 714, pp. 3–8. Springer, Cham (2017). https://doi.org/10.1007/978-3-319-58753-0_1
25. Khan, S., Chang, R.: Anatomy of the vestibular system: a review. Neuro Rehabil. **32**(3), 437–443 (2013)
26. Dennison, M.S., Wisti, A.Z., D'Zmura, M.: Use of physiological signals to predict cybersickness. Displays **44**, 42–52 (2016)
27. Llorach, G., Evans, A., Blat, J.: Simulator sickness and presence using HMDs: comparing use of a game controller and a position estimation system. In: Proceedings of the 20th ACM Symposium on Virtual Reality Software and Technology, pp. 137–140 (2014)
28. Stanney, K.M., Hale, K.S., Nahmens, I., Kennedy, R.S.: What to expect from immersive virtual environment exposure: influences of gender, body mass index, and past experience. Hum. Factors **45**(3), 504–520 (2003)
29. Zielasko, D.: Subject 001—a detailed self-report of virtual reality induced sickness. In: Proceedings—2021 IEEE Conference on Virtual Reality and 3D User Interfaces Abstracts and Workshops, VRW 2021, pp. 165–168 (2021)
30. MacArthur, C., Grinberg, A., Harley, D., Hancock, M.: You're making me sick: a systematic review of how virtual reality research considers gender and cybersickness. In: Proceedings of the 2021 CHI Conference on Human Factors in Computing Systems, pp. 1–15 (2021)

31. Howard, M.C., Van Zandt, E.C.: A meta-analysis of the virtual reality problem: unequal effects of virtual reality sickness across individual differences. Virtual Reality **1**(5), 1–26 (2021)
32. Stanney, K., Fidopiastis, C., Foster, L.: Virtual reality is sexist: but it does not have to be. Front. Robot. AI **7**(4) (2020)
33. Witkin, A.H., Lewis, H.B., Hertzman, M., Machover, K., Bretnall, P., Wapner, S.: Personality Through Perception: An Experimental and Clinical Study, vol. XXVI. Harper, Oxford (1954)
34. Bendall, R.C.A., Galpin, A., Marrow, L.P., Cassidy, S.: Cognitive style: time to experiment. Front. Psychol. **7** (2016)
35. Keshavarz, B., Speck, M., Haycock, B., Berti, S.: Effect of different display types on vection and its interaction with motion direction and field dependence. I-Perception **8**(3), 1–18 (2016)
36. Biocca, F.: Will simulation sickness slow down the diffusion of virtual environment technology? Presence: Teleoper. Virtual Environ. **1**(3), 334–343 (1992)
37. Mittelstaedt, J.M.: Individual predictors of the susceptibility for motion-related sickness: a systematic review. J. Vestibular Res.: Equilibr. Orient. **30**(3), 165–193 (2020)
38. Witkin, H.A., Asch, S.E.: Studies in space orientation. III. Perception of the upright in the absence of a visual field. J. Exp. Psychol. **38**(5), 603–614 (1948)
39. Witkin, H.A., Asch, S.E.: Studies in space orientation. IV. Further experiments on perception of the upright with displaced visual fields. J. Exp. Psychol. **38**(6), 762–782 (1948)
40. Witkin, H.A.: Psychological differentiation and forms of pathology. J. Abnorm. Psychol. **70**(5), 317–336 (1965)
41. Deich, R.F., Hodges, P.M.: Motion sickness, field dependence, and levels of development. Perceptual Motor Skills **36**(3_suppl), 1115–1120 (1973)
42. Witkin, H.A.: The nature and importance of individual differences in perception. J. Pers. **18**(2), 145–170 (1949)
43. Gottschaldt, K.: Über den Einfluß der Erfahrung auf die Wahrnehmung von Figuren - I. Über den Einfluß gehäufter Einprägung von Figuren auf ihre Sichtbarkeit in umfassenden Konfigurationen. Psychol. Forschung **8**(1), 261–317 (1926)
44. Witkin, H.A.: Individual differences in ease of perception of embedded figures. J. Pers. **19**, 1–15 (1950)
45. Lawson, B.D., Proietti, P., Burov, O.: Individual factors impacting cybersickness. In: T. N. S. a. T. Organization (ed.) Guidelines for Mitigating Cybersickness in Virtual Reality Systems, pp. 5.1–5.24 (2021)
46. Zupanic, M.: Aspekte der Prädiktion und Prophylaxe der Simulatorkrankheit. Diplomarbeit an den Fachbereichen Psychologie und Sicherheitstechnik der Universität Wuppertal, 5/93 (1993)
47. Citroën International. https://www.citroen.com/en/Highlight/55/citroenlaunches-the-first-glasses-that-restore-the-taste-fortravel-seetroen. Accessed 10 Nov 2023
48. Brainard, A., Gresham, C.: Prevention and treatment of motion sickness. https://www.aafp.org/afp/2014/0701/p41.html. Accessed 10 Nov 2023
49. Citroën Homepage. https://lifestyle.citroen.de/seetroen-brille-inspired-by-you.html. Accessed 10 Nov 2023
50. Boarding Glasses. https://boardingglasses.com/en. Accessed 10 Nov 2023
51. Rebenitsch, L., Owen, C.: Individual variation in susceptibility to cybersickness. In: UIST 2014 - Proceedings of the 27th Annual ACM Symposium on User Interface Software and Technology, pp. 309–318 (2014). https://doi.org/10.1145/2642918.2647394
52. Reason, J.T.: Relations between motion sickness susceptibility, the spiral after-effect and loudness estimation. Br. J. Psychol. **59**(4), 385–393 (1968)
53. Golding, J.F.: Motion sickness susceptibility questionnaire revised and its relationship to other forms of sickness. Brain Res. Bull. **47**(5), 507–516 (1998)

54. Kennedy, R.S., Lane, N.E., Berbaum, K.S., Lilienthal, M.G.: Simulator sickness questionnaire: an enhanced method for quantifying simulator sickness. Int. J. Aviat. Psychol. **3**(3), 203–220 (1993)
55. Sevinc, V., Berkman, M.I.: Psychometric evaluation of simulator sickness questionnaire and its variants as a measure of cybersickness in consumer virtual environments. Appl. Ergon. **82**, 102958 (2020)
56. Stone III, W.B.: Psychometric evaluation of the Simulator Sickness Questionnaire as a measure of cybersickness. Iowa State University, Ames (2017)
57. Kim, H.K., Park, J., Choi, Y., Choe, M.: Virtual reality sickness questionnaire (VRSQ): motion sickness measurement index in a virtual reality environment. Appl. Ergon. **69**, 66–73 (2018)
58. Keshavarz, B., Murovec, B., Mohanathas, N., Golding, J.F.: The visually induced motion sickness susceptibility questionnaire (VIMSSQ): estimating individual susceptibility to motion sickness-like symptoms when using visual devices. Hum. Factors 1–17 (2021)
59. Bock, O.L., Oman, C.M.: Dynamics of subjective discomfort in motion sickness as measured with a magnitude estimation method. Aviat. Space Environ. Med. **53**(8), 773–777 (1982)
60. McHugh, N., Jung, S., Hoermann, S., Lindeman, R.W.: Investigating a physical dial as a measurement tool for cybersickness in virtual reality. In: Proceedings of the 25th ACM Symposium on Virtual Reality Software and Technology (VRST 2019), pp. 1–5. ACM, New York (2019)
61. Kreuzpointner, L., Lukesch, H., Horn, W.: Leistungsprüfsystem 2. LPS-2. Hogrefe, Göttingen (2013)
62. Döring, N.: Forschungsmethoden und Evaluation in den Sozial- und Humanwissenschaften, 6th edn. Springer, Berlin (2023)
63. Islam R., et al.: Automatic detection and prediction of cybersickness severity using deep neural networks from user's physiological signals. In: IEEE International Symposium on Mixed and Augmented Reality (ISMAR), pp. 400–411, Porto de Galinhas, Brazil (2020)
64. Madary, M.; Metzinger, T.K.: Real virtuality: a code of ethical conduct. Recommendations for good scientific practice and the consumers of VR technology. Front. Robot. AI **3**(3) (2016)
65. Bimberg, P., Weissker, T., Kulik, A.: On the usage of the simulator sickness questionnaire for virtual reality research. In: 2020 IEEE Conference on Virtual Reality and 3D User Interfaces, Abstracts and Workshops (VRW), pp. 464–467 (2020)
66. Litleskare, S.: The relationship between postural stability and cybersickness: It's complicated – an experimental trial assessing practical implications of cybersickness etiology. Physiol. Behav. **236**, 113422 (2021)
67. Carroll, J.B.: Human Cognitive Abilities: A Survey of Factor-Analytic Studies. Cambridge University Press, Cambridge (1993)
68. Giancola, M., Palmiero, M., Piccardi, L., D'Amico, S.: The Relationships between cognitive styles and creativity: the role of field dependence-independence on visual creative production. Behav. Sci. **12**, 212 (2022)
69. Lienert, G.A., Raatz, U.: Testaufbau und Testanalyse. Beltz, Weinheim (1994)
70. Boorsboom, D., Mellenbergh, G.J., van Heerden, J.: The concept of validity. Psychol. Rev. **111**, 1061–1071 (2004)
71. Curry, C.: Cybersickness in virtual reality: examining the influence of the virtual environments on sex susceptibility. In: SIGGRAPH Asia 2019 Doctoral Consortium (SA 2019), vol. 7, pp. 1–4. ACM, New York (2019)

72. Harth, J., et al.: Different types of users, different types of immersion. a user study on interaction design and immersion in consumer virtual reality. IEEE Consum. Electron. Mag. **7**(4), 36–43 (2018)
73. Grassini, S., Laumann, K.: Are modern head-mounted displays sexist? A systematic review on gender differences in HMD-mediated virtual reality. Front. Psychol. **7**(11), 1604 (2020)
74. Picot, E., Wright, C.I.: Motion sickness – newly engineering glasses may offer a solution. J. Appl. Mech. Eng. **5**(2), 197 (2016)

Temporal and Flexible Data Warehouses

Redha Benhissen(✉), Fadila Bentayeb, and Omar Boussaid

ERIC Laboratory, University of Lyon 2, 5 Av. Pierre Mendès, Bron, France
redha.benhissen@univ-lyon2.fr

Abstract. The present trend among companies involves a significant revamp of their data architecture, aiming to streamline data processes and phase out outdated systems. The advent of big data has profoundly influenced businesses, empowering them to adeptly manage and analyze vast datasets. In the realm of business intelligence, particularly in decision-making, data warehouses play a crucial role, leveraging OLAP technology for the efficient analysis of structured data. Constructed by amalgamating data from diverse sources, a data warehouse faces the challenge of accommodating big data-comprising unstructured, semi-structured, or structured data from myriad sources-where alterations in content and structure are frequent. To address this, our paper introduces a temporal multidimensional model utilizing a graph formalism for multi-version data warehouses, adept at assimilating changes from data sources. This approach relies on multi-version evolution for schema modifications and employs bi-temporal labeling for entities and their relationships to capture data evolution. Our proposal enhances data warehouse evolution flexibility, broadening analysis possibilities within the decision support system, and enabling adaptable temporal queries to yield consistent results. Building upon our prior work [6], where we presented the GAMM model emphasizing evolutionary data treatment, including dimensional changes, this study expands on the temporal labeling principle in our approach. It delves into various functions governing the evolution of temporal data, offering illustrative examples. We validate our approach through a case study demonstrating temporal queries and conduct runtime performance tests on graph data warehouses.

Keywords: Temporal data warehouse · Multidimensional model · Data evolution · Slowly changing dimension · Temporal query · Graph-based database · NoSQL

1 Introduction

The architectures of data warehouses (DWs) enable the storage of data extracted from various and heterogeneous sources in a cohesive and integrated manner. This provides decision-makers with a better understanding of their environments and adequate support for strategic decision-making [18,21]. Since the inception of DWs, the warehousing approach has become a significant research field, with many challenges, particularly concerning the evolutionary aspect of DWs.

The proliferation and diversification of data sources, facilitated by the advent of big data, lead to changes in the content and structure of DWs. The schema, designed to meet predefined analysis needs, becomes costly to modify if those needs change.

O. Gusikhin et al. (Eds.): DATA 2023, CCIS 2105, pp. 25–49, 2024.
https://doi.org/10.1007/978-3-031-68919-2_2

The classical multidimensional model, relying on the star schema and its variants, faces limitations in accommodating change, and its evolution is complex. Our prior work introduced a flexible multidimensional model for big data, the graph-based agile multidimensional model (GAMM). GAMM extends the classical multidimensional model to support chronological evolution at the conceptual level and the evolution of the graph structure at the logical and physical levels [4,5]. GAMM facilitates schema evolution in a data warehouse by creating a new version of the schema at each evolution using evolution functions.

For an accurate representation of the analysis context, periodic and regular data refreshing through ETL (extract, transform, load) processes is crucial. However, once a data warehouse is operational, changes in the dimensions, especially in hierarchical relationships or descriptor attributes, may occur. Properly managing these changes is vital for data warehousing system consistency. Our previous work [6] addressed the evolutionary treatment of data in the GAMM model, including dimensional changes. In this work, we expand on the previous work by presenting the temporal labeling principle in our approach and detailing various functions defined for the evolution of temporal data, illustrated with examples. We also depicted the various multidimensional concepts using graphs to enhance understanding of our approach.

Changes in dimensions necessitate the study of temporality, where a piece of information/relationship is true during a specific period. Temporal data warehouses (TDWs), inspired by the principles and rules of temporal databases, facilitate managing this temporality notion, offering coherent analyses despite changes. In this article, we propose a temporal approach for multi-version DWs based on a graph database to retain the evolutionary history of data, including changes in dimensions, allowing flexible temporal queries for consistent results.

The rest of the article is organized as follows: A motivating example is presented in Sect. 2 followed by related work on data evolution and temporal approaches in a DW context in Sect. 3. Section 4 introduces our concept of a temporal multidimensional model for multi-version DWs based on a graph formalism and we have explained the principle of time management in the. In Sect. 5, we detail various functions defined for the evolution of temporal data. Temporal queries are provided in Sect. 6. In Sect. 7, a use case based on the Star Schema Benchmark (SSB) dataset is presented for functional validation and studying runtime performance. Finally, conclusions and future research topics are outlined.

2 Example of Data Evolution in Data Warehouse

In addition to inserting new entities (facts and dimensions) into the data warehouse, existing data may undergo modifications through update operations. These modifications involve changing existing attributes and/or altering the relationships between hierarchies, necessitating updates to keys. If not handled correctly, these update operations can impact the analyses generated by the data warehouse. We employ the dimensional fact model (DFM) formalism [14] to represent the conceptual schemas in our example (Fig. 1). The example comprises a fact, denoted as SALES, and three dimensions: PRODUCT, CUSTOMER, and DATE. For consistency, we will maintain this schema configuration as a running example throughout the presentation of our proposed approach.

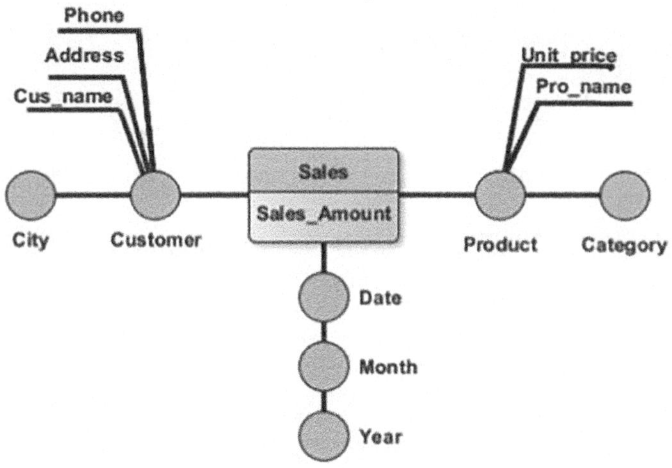

Fig. 1. Multidimensional model schema for retail sales using DFM formalism.

The PRODUCT dimension includes the attributes *Product_Name* and *Unit_Price*, as well as the hierarchical level CATEGORY. This dimension, therefore, features a single hierarchy on the analysis axis (PRODUCT, CATEGORY). The CUSTOMER dimension comprises the attributes *Customer_Name*, *Address*, and *Phone*, along with the hierarchical level CITY. The DATE dimension has two hierarchical levels, namely MONTH and YEAR. The SALES fact is described by the measure *Sales_Amount*.

Below, we provide an example of the temporal evolution of data in a data warehouse (DW) to illustrate a case involving changes in dimensions. It's important to note that the representation of instances in this example adheres to the principles of temporal databases, allowing data to possess advanced chronological precision. Information is considered valid during a specific lifespan characterized by a From_Date (FD) and a To_Date (TD). Entities are characterized by a Valid_Time (VT) and/or a Transaction_Time (TT) in a mono- or bi-temporal approach [16]. This temporal notion enhances the analysis level to better represent the real world, especially in managing the temporality of aggregation links. The instances of the CUSTOMER dimension, as shown in Table 1, indicate that the customer *Mary Saveley* initially lived in the city of *Paris* from 01/01/2020 to 31/07/2020, and later moved to *Lyon*, where she has resided since 01/08/2020. Similarly, the instances of the PRODUCT dimension, displayed in Table 2, demonstrate that the product *Mozzarella* was initially categorized as *Fresh* from 01/01/2020 to 31/03/2020 and was subsequently reclassified as *Dairy* on 01/04/2020.

Table 4 displays instances of the SALES fact, describing the *Sales_Amount* for the customer *Mary Saveley* and the product *Mozzarella*. These sales amounts represent the

Table 1. Dimension CUSTOMER.

Surrogate_Key	Customer_ID	Customer_Name	City_ID	From_Date	To_Date
SkCust001	Cust001	Mary Saveley	City001	01/01/2020	31/07/2020
SkCust002	Cust001	Mary Saveley	City002	01/08/2020	31/12/9999

Table 2. Dimension PRODUCT.

Surrogate_Key	Product_ID	Product_Name	Category_ID	From_Date	To_Date
SkProd001	Prod001	Mozzarella	Categ001	01/01/2020	31/03/2020
SkProd002	Prod001	Mozzarella	Categ002	01/04/2020	31/12/9999

Table 3. Levels CITY and CATEGORY.

City_ID	**City_Name**
City001	Paris
City002	Lyon

Category_ID	**Category_Name**
Categ001	Fresh
Categ002	Dairy

Table 4. Fact SALES.

Sale_ID	Customer_ID	Product_ID	Order_Date	Sales_Amount
Sale001	Cust001	Prod001	01/03/2020	1500
Sale002	Cust001	Prod001	01/05/2020	2200
Sale003	Cust001	Prod001	01/09/2020	1800
Sale004	Cust001	Prod001	01/11/2020	2000

customer's purchases both when she lived in Paris and when she resided in Lyon. Furthermore, these amounts also represent the sales of the product *Mozzarella* during its classification under the *Fresh* category and its subsequent reclassification under the *Dairy* category.

Kimball and Ross addressed the issue of changes in dimensions and proposed processing techniques based on the speed of these changes [22]. For slowly changing dimensions (SCD), the following types of techniques are suggested; (i) Maintain the original value of the attribute, ensuring that facts are always grouped by this original value (Type 0). (ii) Replace the old value of the attribute with a new one, associating the facts with the current value of the attribute (Type 1). (iii) Add a new dimension row with a new value for the attribute/foreign key, considering the temporal aspect. This ensures that facts are associated with the attribute value based on the time span of its veracity, delimited by a start date and an end date (Type 2). (iv) Add a new column to preserve both the current and previous values of the attribute (Type 3). For fast-changing dimensions, the proposed technique involves adding a mini-dimension. This way, frequently analyzed or rapidly changing attributes are split into a separate dimension. Other hybrid techniques have also been proposed, which combine some or all of the different techniques mentioned above to meet the requirements of historical attribute preservation and reporting.

In most cases, the Type 2 technique, which involves adding a new row with the new value while considering the time interval of data veracity, is the most representative of the real world and provides consistent analyses. The Type 0 technique (maintaining the original value) and the Type 1 technique (replacing it with the new value) lead to information losses. Similarly, the Type 3 technique of adding a new column is impractical for continuous application due to schema modifications with each addition. Regarding the technique of adding a mini-dimension, it is intended for specific cases and does not offer a general solution to the issue of changes in dimensions.

3 Related Work

There are numerous research works in the literature that have explored various aspects of temporal data warehouses, covering topics such as types of temporality, conceptual and logical levels, data evolution, changes in dimensions, delayed measurements, implementation approaches, temporal queries, and aggregation in temporal relationships [12, 17]. Notably, [7] presented a temporal approach based on suppressing the time dimension, replacing it with temporal labels such as VT or TT (or both for a bi-temporal approach) at the instance level. In the realm of temporal OLAP queries, [23] introduced a temporal multidimensional model. [15] proposed alternative design solutions to support meaningful historical analysis in the presence of late measurements. Additionally, [11] classified queries based on input and output attributes, studying the performance of Type 2 and hybrid Slowly Changing Dimensions (SCDs). [25] presented an approach managing dimension data and tracking retroactive and proactive updates in a bi-temporal DW using both VT and TT. [13] introduced a logical modelling approach for Temporal Data Warehouses (TDWs) based on the temporal starnest schema. [24] utilized temporal database features, including VT state tables, to solve the SCD problem in data warehouses, particularly in SQL. Furthermore, [1,2] presented a logical model and querying technique for temporal data warehouses. While these studies have offered various solutions for managing data evolution in data warehouses and addressing SCD issues, they primarily rely on the entity-relationship (ER) model, which limits the evolution of multidimensional models, especially concerning schema evolution [4,5]. In response, this paper proposes an alternative approach for temporal data management in DWs using graph-oriented NoSQL databases, building on promising results obtained by models like those proposed by [9, 10]. This approach provides a comprehensive solution to the SCD problem and allows for temporal queries with consistent results.

4 Temporal Multidimensional Model Based on a Graph Formalism

To introduce our approach, let's provide a concise overview of our previously proposed model known as the graph-based agile multidimensional model (GAMM) [5]. The focus of this model lies in the evolution of the schema. However, in this paper, our emphasis is on the temporal evolution of the data, and we extend the formalism of the model to accommodate this evolution, particularly addressing changes in dimensions and schema evolution.

In our context, agility in a multidimensional model denotes its organizational capability to construct scalable data warehouses, prioritizing the objectives of business teams. Agility entails responsiveness to new business goals and the seamless integration of new data sources in a flexible and incremental manner while retaining all previous builds. This flexibility enhances the business value delivered to decision-makers, enabling better environmental management.

GAMM serves as a flexible solution for schema and data evolution within data warehouses. This model empowers designers to integrate new data sources and cater to emerging user requirements, thereby enriching the analytical capabilities of the data warehouse. GAMM relies on a multi-version scalable schema model and a data warehouse stored in a unified global graph database, as illustrated in Fig. 2.

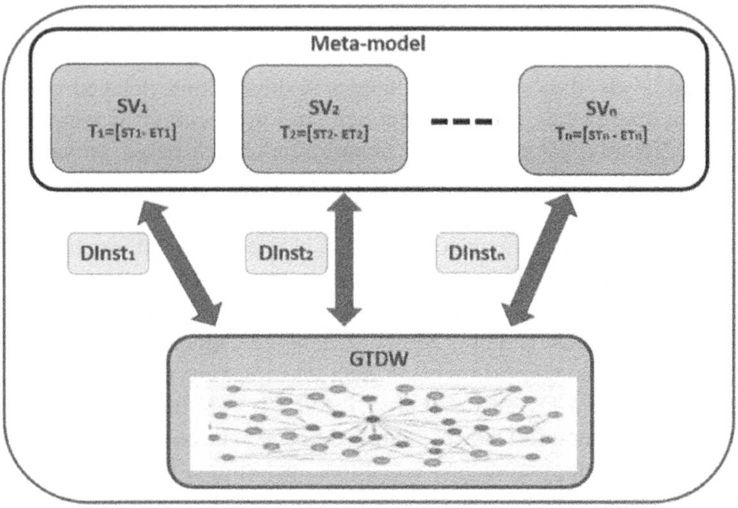

Fig. 2. Architecture of GAMM.

Each schema version (SV) is valid for a specific time period (T), defined by a Starting_Time (ST) and an Ending_Time (ET). It corresponds to a data instance (DInst) extracted from the graph-based temporal data warehouse (GTDW). To manage schema versions with sequential validity periods, we implement a meta-model. The subsequent subsections will delve into the formalization of our approach (Subsect. 4.2), the concept of the GTDW (Subsect. 4.3), and the transformation rules (Subsect. 4.4).

4.1 Principle of Temporal Labeling in the GAMM Model

The fundamental principle of temporal databases is to associate a temporal value with each record through a timestamping mechanism. This mechanism allows tracking changes and the validity period of data over time, enabling queries based on this temporal information [8]. Data timestamping is a mechanism that involves associating a date (and possibly a time) with a piece of data. Its primary objective is to mark a reference moment when an operation was performed or to record the validity period of the data. Timestamping can be either punctual, indicating a specific moment, or a time interval, covering a specific duration. The choice between the two depends on the type of data and user requirements [8]. In a punctual timestamping, each record in the database is associated with a "timestamp" representing a specific moment, such as Valid_Time (VT) for validity date [20], Transaction_Time or Load_Time (TT), Decision_Time (DT) [19], etc. Punctual timestamps are useful for recording specific real-world events and are commonly used in databases for tracking, logging, and decision-making operations. Valid_Time represents the moment when data is true concerning reality, while Transaction_Time represents the moment when data is stored in a database [19,20]. The bitemporal data model supports both validity time and transaction time [8].

To preserve the evolutionary history of data, including changes in dimensions, and in line with the principles of temporal databases, we applied a VT-type temporal labeling to nodes of the "Fact" type. Additionally, FD and TD-type temporal labeling were applied to relationships at the dimension instance level, identifying the validity period of these relationships. Through these temporal labels, we ensure consistent analyses and concordant results, even in the presence of changes in the data. It is important to note that in punctual timestamping, VT represents the moment when data is true concerning reality, while TT represents the moment when this data is stored in a database. On the other hand, in interval timestamping, FD represents the start time of the validity period of the data, while TD represents the end time of this temporal validity period. The labeling we use allows the identification of different versions of the schema and data instances corresponding to each version. This enables us to perform homogeneous analyses and obtain consistent results, even in the presence of changes in the data, especially those in the dimensions.

4.2 Formalisation of the Approach

On the conceptual level, the approach represents an extension of classical multidimensional modelling based on the *fact, measure, dimension, level* and *hierarchy* concepts. Indeed, due to the evolution of the schema and data over time and for historical purposes, the temporal concept has been introduced according to the following definitions.

Definition 1. GAMM is represented as follows:

$$GAMM(t) = \{F, D, FAssoc[F, D](t)\}.$$

$GAMM(t)$ represents the schema version at a time t.
$t \in T = [ST, ET]$ represents the period of validity of the schema version, where ST is the *Starting_Time* of the version and ET is the *Ending_Time* of the version.
$F = \{f_i(t)\}, i \in [1, *]$, represents the set of facts at the moment t.
$f_i(t)$ represents the fact f_i at a time t.
$D = \{d_j(t)\}, j \in [1, *]$, represents the set of dimensions according to which $f_i(t)$ is analysable at a time t.
$d_j(t)$ represents the dimension d_j at a time t.
$FAssoc[F, D](t) : f_i(t) \Longrightarrow \{d_j(t), ST, ET\}$, where $j \in [1, *]$, represents the association function of the set of dimensions $\{d_j(t)\}$ and the fact $f_i(t)$ at a time t.

Example:

$$GAMM(t_0) = \{\{Sales\}, \{Customer, Product\}, \{Sales \Longrightarrow Customer, Product\}\}$$

and $ST_0 =< t_0 < ET_0$.

Definition 2. A measure is an indicator allowing the analysis of the business subject represented by the fact; it is numerical and aggregable and is defined as follows:

$$M(t) = \{M\_label, I_k^m\}.$$

M_label represents the measure identifier.
I_k^m, $k \in [1, *]$, represents the set of instances of measure M.

Example:

$$M(t_o) = \{Sales\_Amount, I_0^m : \{1500\}\}.$$

Definition 3. A fact represents a subject analysed by GAMM. It is defined as follows:

$$F(t) = \{F\_label, [M], MAssoc[F, M](t), [I^f]\}.$$

F_label represents the fact name.
$[M] = \{m_k(t)\}$, $k \in [1, *]$, represents the set of measures associated with the fact at a time t.
$m_k(t)$ represents a measure m_k at an instant t.
$MAssoc[F, M](t) : f_i(t) \Longrightarrow \{m_k(t), ST, ET\}$ represents an association function of the set of measures $\{m_k(t)\}$ and the fact $f_i(t)$ at a time t, where ST is the *Starting_Time* and ET is the *Ending_Time*.
$[I^f] = \{i_l^f\}$, $l \in [1, *]$, represents the set of instances of fact F. Each instance $i^f = \{[I_k^m], VT, TT\}$.
$TT = Transaction\_Time$ represents the time point at which a fact instance is stored in the model.
$VT = Valid\_Time$ represents the time point at which a fact instance is true in relation to reality.
Example:

$$F(t_o) = \{Sales, M_o, MAssoc[F, M](t_o), I_0^f\}.$$

$M_o : \{Sales\_Amount\}$.
$MAssoc[F, M](t_o) : \{Sales \Longrightarrow Sales\_Amount\}$.
i_0^f: $\{I_0^m : \{1500\}$, 05/01/2020, 15/06/2020$\}$.

Definition 4. An attribute is an element of description of a dimension or a hierarchical level to which it is associated. It is defined as follows:

$$A(t) = \{A\_Label, Attribute, [I^a]\}.$$

A_Label represents the attribute name.
Attribute represents the value of the description attribute.
$[I^a] = \{i_k^a\}$, $k \in [1, *]$, represents the set of instances of attribute A. Each instance $i^a = \{value, VT, TT\}$.
$VT = Valid\_Time$ represents the time point at which an attribute is true with respect to reality.
$TT = Transaction\_Time$ represents the time point at which an attribute is stored in the model.

Example.

$$A(t_0) = \{Product\_Name, name,$$
$$i_0^a : \{Mozzarella, 05/01/2020, 15/06/2020\}\}.$$

Definition 5. A level represents the degree of detail of an analysis perspective according to a given hierarchy. It is defined as follows:

$$L(t) = \{L\_Label, L\_ID, A, [I^l], LAssoc[L, A](t), Rel[[I^l], [I^a]]\}.$$

L_Label represents the level name.
L_ID represents the level identifier.
$A = \{a_i(t)\}, i \in [1, *]$, represents the set of attributes associated with the level at a time t.
$[I^l] = \{i_k^l\}$, $k \in [1, *]$, represents the set of instances of level L. Each instance $i^l = \{id, VT, TT\}$.
$VT = Valid\_Time$ represents the time point at which a level is true compared to reality.
$TT = Transaction\_Time$ represents the time point at which a level is stored in the model.
$LAssoc[L, A](t) : L_j(t) \Longrightarrow \{a_i(t), ST, ET\}$ represents the association function of the set of attributes $\{a_i(t)\}$ and the level $L_j(t)$ at a time t, where ST is the *Starting_Time*, ET is the *Ending_Time* and $j \in [0, *]$. $Rel[[I^l], [I^a]]$ represents the relationships between instances of levels [L] and attributes [A], where $Rel[[i_m^l], [i_n^a]]$ = {Relation_Label, From_Date, To_Date}.
$Relation\_Label$ represents the relationship name.
$From\_Date$ represents the starting date of the relationship.
To_Date represents the ending date of the relationship.

Example.

$$L(t_0) = \{Category, Category\_Id, A_0, i_0^l, LAssoc[L, A_0](t_0), Rel[[I_0^l], [I_0^a]]\}.$$

$A_0 : \{Category\_Name\}$.
$i_0^l : \{Categ001, 05/01/2020, 15/06/2020\}$
$LAssoc[L, A_0](t_0) : \{Category \Longrightarrow Category\_Name\}$.
$Rel[[I_0^l], [I_0^a]] = \{Category\_To\_Name, 05/01/2020,$
$31/12/9999\}$
$i_0^a : \{Fresh, 05/01/2020, 15/06/2020\}$.

Definition 6. A dimension is an axis of analysis according to which the business subject is analysed. It determines the level of detail of the measures and is defined as follows:

$$D(t) = \{D\_Label, D\_ID, [A], [L], [I^d], DAssoc_a[D, A](t),$$
$$DAssoc_l[D, L](t), Rel[[I^d], [I^a]], Rel[[I^d], [I^l]]\}.$$

D_Label represents the dimension name.
D_ID represents the dimension identifier.
$[A] = \{a_i(t)\}, i \in [1, *]$, represents the set of attributes associated with the dimension at a time t.
$a_j(t)$ represents the attribute a_j at a time t.
$[L] = \{l_k(t)\}, k \in [0, *]$, represents the set of levels associated with the dimension at a time t.
$[I^d] = \{i^d_k\}$, $k \in [1, *]$, represents the set of instances of dimension D. Each instance $i^d = \{id, VT, TT\}$.
$VT = Valid\_Time$ represents the time point at which a dimension is true in relation to reality.
$TT = Transaction\_Time$ represents the time point at which a dimension is stored in the model.
$DAssoc_a[D, A](t) : d_j(t) \Longrightarrow \{a_i(t), ST, ET\}$ represents the association function of the set of attributes $\{a_i(t)\}$ and the dimension $d_j(t)$ at a time t, where ST is the *Starting_Time* and ET is the *Ending_Time*.
$Rel[[I^d], [I^a]]$ represents the relationships between instances of dimension [D] and attributes [A], where $Rel[[i^d_m], [i^a_n]]$ = {Relation_Label, From_Date, To_Date}.
$Relation\_Label$ represents the relationship name.
$From\_Date$ represents the starting date of the relationship.
To_Date represents the ending date of the relationship.
$DAssoc_l[D, L](t) : (d_j(t) \Longrightarrow \{l_k(t), ST, ET\}$ represents the association function of the set of levels $\{l_k(t)\}$ and the dimension $d_j(t)$ at a time t, where ST is the *Starting_Time*, ET is the *Ending_Time* and $j \in [0, *]$.
$Rel[[I^d], [I^l]]$ represents the relationships between instances of dimensions [D] and levels [L], where $Rel[[i^d_m], [i^l_n]]$ = {Relation_Label, From_Date, To_Date}.
$Relation\_Label$ represents the relationship name.
$From\_Date$ represents the starting date of the relationship.
To_Date represents the ending date of the relationship.

Example.

$$D(t_0) = \{Product, Product\_ID, A_0, L_0, i^d_0, DAssoc_a[D, A](t_0), DAssoc_l[D, L](t_0),$$
$$Rel[[I^d_0], [I^a_0]], Rel[[I^d_0], [I^l_0]]\}.$$

$A_0 : \{Product\_Name\}$.
$L_0 : \{Category\}$.
$i^d_0 : \{Prod001, 05/01/2020, 15/06/2020\}$
$DAssoc[D, A](t_o) : \{Product \Longrightarrow Product\_Name\}$.
$Rel[[I^d_0], [I^a_0]] = \{Product\_To\_Name, 05/01/2020, 31/12/9999\}$
$i^a_0 : \{Mozzarella, 05/01/2020, 15/06/2020\}$
$DAssoc_l[D, L](t_0) : \{Product \Longrightarrow Category\}$.
$Rel[[I^d_0], [I^l_0]] = \{Product\_To\_Category, 05/01/2020, 31/12/9999\}$
$i^l_0 : \{Prod001, 05/01/2020, 15/06/2020\}$.

Definition 7. A hierarchy is a projection of analysis by level along the axis defined by the dimension. It is organised from the finest to the coarsest granularity level, thus offering analysis possibilities for ascending groupings through *roll-up* and descending groupings through *drill-down*. The hierarchy is defined as follows:

$$H(t) = \{L, R_h[L_j, L_k](t)\}.$$

$L = \{l_i(t)\}, i \in [1, *]$, represents the set of aggregation levels constituting a hierarchy H(t) at a time t.
$R_h[l_j, l_k](t)$ represents the aggregation function between the different levels $\{l_i(t)\}$ constituting a hierarchy $H(t)$ at a time t, with $j \in [1, *], k \in [1, *]$ and $j \neq k$.
Assume that $D_i(t) \in D$ with $i \in [1, *]$, $\forall\ H_j(t) \in H$ with $j \in [1, *]$, and $L_1^{h_j}(t)$ is directly related to $D_i(t)$ so that

$$D_i(t) \prec L_1^{h_j}(t) \prec L_2^{h_j}(t) \prec ... \prec L_k^{h_j}(t) \prec All.$$

$D_i(t)$ represents the finest level of aggregation of the analysis axis.
$L_k^{h_j}(t)$ represents the coarsest level of aggregation in the hierarchy h_j.
All represents the global aggregation level of the dimension.

Example: Analysis projection by COUNTRY and CITY for the CUSTOMER dimension can be performed as follows:

$$D(t) = (Customer) \prec L_1^{h_j}(t) = (City) \prec L_2^{h_j}(t) = (Country) \prec All.$$

This formalism provides flexibility in the evolution of the model, both at the schema level and at the level of data instances, while preserving the history of these evolutions. For the purpose of studying data evolution and clarity in explanation, we assume that the schema does not change. The data instances in the Graph-Based Temporal Data Warehouse (GTDW) evolve incrementally and without redundancy. All data instances from previous schemas remain available for consultation.

In cases where the relationship between entities of a dimension/level and a lower level changes, a new link to the corresponding instance in this level is created. This new link is labeled with an FD label set to the creation date and a TD label set to 31/12/9999. Simultaneously, the TD of the old relationship is set to the same creation date. Consequently, each relationship is valid during a specific time interval. The same principle is applied in the case of changing descriptor attributes, as these are also represented by external nodes in relation to the business concept. Queries can be parameterized with these time labels to obtain results that accurately represent the real state of the Data Warehouse and take into account any changes in the dimensions. This approach ensures that historical data instances are preserved and accessible for analysis, providing a comprehensive view of the DW's evolutionary journey.

4.3 Graph Temporal Data Warehouse: A Paradigm Shift

The Graph Temporal Data Warehouse (GTDW) represents a paradigm shift from the classical DW by incorporating the temporal concept into aggregation relationships, driven by the dynamic nature of the underlying data. This evolution is marked by the separation of business concepts from their descriptors (attributes), enabling each entity to undergo independent evolution. The choice of a graph formalization and its implementation in a graph database was a strategic decision to overcome the limitations associated with the Entity-Relationship model. The graph formalization provides increased flexibility, especially concerning model evolution [3]. In addition to the inherent representational quality of interconnected data and the utilization of information-carrying links, especially in the context of temporal aspects in aggregation links, the adoption of graphs was motivated by the absence of integrity constraints, the lack of a pre-established schema, and the ability to represent each value of a tuple (or the entire tuple) using a graph node.

This graph-based implementation diverges from traditional relational tables, where schemas consist of fixed horizontal lines and vertical columns. Any addition or removal of a column in this structure affects the entire schema. In contrast, a (NoSQL) graph database (GDB) was employed to embody the GTDW based on our formalization. GDBs represent data in the form of a graph (vertices/edges) using physical pointers between nodes, eliminating the need for joins in queries, especially beneficial in a big data context. GDBs also lack strict data types and allow the integration of information into the relationships between data. Moreover, the adoption of graph concepts facilitates advanced analysis of data representation, enabling online analysis for the generation of explanatory and predictive models. This approach embraces the dynamic nature of data, providing a robust foundation for evolving and adapting to changing business requirements.

4.4 Transformation Rules from Classical Approch to Graph Approach

According to the formalisation presented in Subsect. 4.2 and the characteristics of the graph databases, the business concepts, as well as the descriptors, will be represented by nodes, and the relationships between these concepts will be represented by edges (Fig. 3). According to the formalisation of our approach, we have established the following rules for moving from a classical multidimensional model to a graph multidimensional model:

1. Each tuple of a fact is represented by its own node.
2. Each (business/descriptor) value of a dimension, level or attribute is represented by its own node.
3. All relationships are represented by edges.
4. The fact nodes contain the measures.
5. The fact nodes are directly related to the dimension nodes.
6. The level nodes are related to a dimension/another level node according to their depth.
7. The attribute nodes are directly related to the dimension/level nodes.
8. The levels constitute hierarchies according to axes of analysis organised from the finest to the coarsest level of aggregation.
9. All edges between the (business/descriptor) values of a dimension, level or attribute have a chronological FD label and TD label to determine the lifespan of the relationship.
10. All fact nodes have a chronological VT label and TT label in accordance with the principles of bi-temporal databases.

Applying the transformation rules to the schema of our ongoing example, we derive the updated schema depicted in Fig. 3. This representation encompasses the three dimensions: PRODUCT, CUSTOMER, and DATE, along with the fact SALES, the measure *Sales_Amount*, and the hierarchical levels CATEGORY (within PRODUCT) and CITY (within CUSTOMER). The nodes within the diagram represent descriptors, encapsulating the details of dimensions and levels.

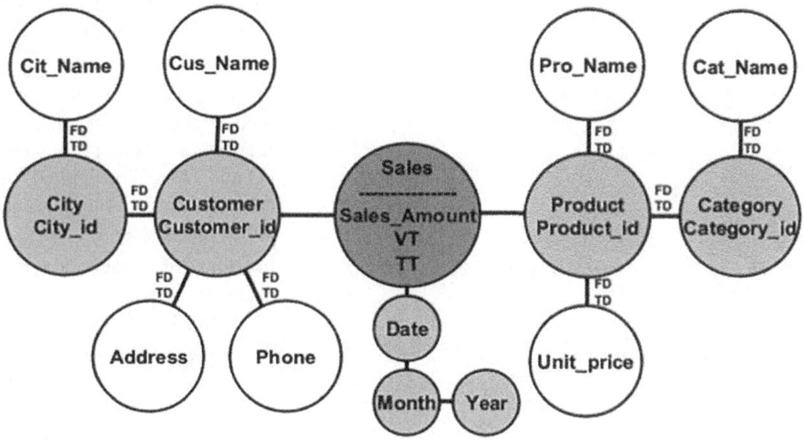

Fig. 3. Logical schema for the GTDW.

This model introduces a clear distinction between business concepts and descriptors (attributes), facilitating the independent evolution of each entity. Basic multidimensional concepts such as facts, dimensions, and hierarchy levels are denoted by nodes, with additional nodes representing descriptors linked to them. Temporal labels, encompassing VT and TT, have been assigned to all entities, aiding in the identification of distinct instances. Furthermore, the FD and TD parameters dictate the durations of relationships within dimensions. This ensures the GTDW can seamlessly handle queries, preventing discrepancies in results arising from changes in dimensions.

5 Data Evolution Functions in GAMM

Data instances in GAMM evolve incrementally and without redundancy. All data instances from previous schemas are available for consultation. The data evolution functions are defined in Table 5 as follows:

The functions presented in Table 5 are used in the context of data evolution. They allow performing initial loading, refreshing, and modification operations on the data while preserving the history of these changes. Temporal labeling of timestamp or interval type will be applied to nodes and relations, following the principles of temporal databases and the formalism of the GAMM model presented in the Sect. 4.

5.1 The « Load_data() » Function

The « Load_data() » function is used for the initial loading of data from data sources into the GAMM model. This operation involves extracting data from sources, transforming if necessary, and loading it into the data warehouse. During the initial data

Table 5. Data Evolution Functions.

Evolution	Evolution Function	Description
Initial Data Loading	*Load_data(Data_sources)*	Initial loading of data from data sources. A punctual timestamp of type VT and TT will be applied to nodes representing facts, while a temporal labeling of interval type FD and TD will be applied to relations between dimension nodes
Dimension Refresh	*Refresh_dimension(Data_sources)*	Loading of new dimension entities (nodes/relations) from data sources. A temporal labeling of interval type FD and TD will be applied to relations between the relevant dimension nodes
Fact Refresh	*Refresh_fact(Data_sources)*	Loading of new fact nodes from data sources and creation of connections with corresponding dimension nodes. A punctual timestamp of type VT and TT will be applied to the relevant nodes
Dimension Modification	*Change_dimension*($\{D, A\}$), *Change_dimension*($\{L, A\}$), *Change_dimension*($\{D, L\}$), *Change_dimension*($\{L, L\}$)	Modification of a description attribute or hierarchical link in dimensions. Creation of a new attribute node and connection with the appropriate node (dimension/level) in case of attribute change, or creation of a new link to the appropriate level in case of hierarchical relation change. A temporal labeling of interval type FD and TD will be applied to relations between dimension nodes, and the TD of the old link will be updated
Fact Modification	*Change_fact*($\{F\}$)	Modification of fact nodes in case of errors in data sources. The punctual timestamp of type VT and TT of the relevant nodes will be maintained

loading, temporal labeling is applied to nodes representing facts and relations between dimension nodes. Specifically:

1. Nodes representing facts are labeled with a punctual timestamp of type VT and TT. The VT attribute takes the actual validity date of the instance from the data sources, while TT takes the date the data is loaded into the data warehouse. This assigns a

temporal value to each fact instance, indicating when it is valid and when it was loaded into the data warehouse.

2. Relations between dimension nodes are labeled with a temporal labeling of interval type represented by the FD and TD attributes. FD takes the start date of the relation from the data sources, while TD takes the date 31/12/9999. These labels indicate the validity period of each relation between dimension values (the date 31/12/9999 means the relation is considered always valid and has not expired yet).

5.2 The « Refresh_dimension() » Function

The « Refresh_dimension » evolution function is used to load new dimension entities from data sources into the GAMM model. This allows adding new dimension instances or new relations between existing dimension nodes. During the loading of these new dimension entities, a temporal labeling of interval type, represented by the FD and TD attributes, is applied to relations between the relevant dimension nodes. Similar to the initial data loading, FD takes the start date of the relation from the data sources, while TD takes the date 31/12/9999. This defines the validity period of these relations, indicating when they are active and when they were loaded into the data warehouse.

5.3 The « Refresh_fact() » Function

The « Refresh_fact» evolution function is used to load new fact nodes from data sources and link them to the corresponding dimension nodes in the GAMM model. This operation adds new fact instances to the data warehouse. During the loading of these new fact nodes, a temporal labeling of punctual timestamp type is applied to the relevant nodes, using the VT and TT attributes. Similar to the initial data loading, VT takes the actual validity date of the instance from the data sources, while TT takes the date the data is loaded into the data warehouse. This assigns a temporal value to each new fact instance, indicating when it is valid and when it was modified.

Additionally, during loading, new fact nodes are linked to the corresponding dimension nodes. These relationships are established based on the matching keys between fact and dimension data (typically, fact refreshing is always preceded by dimension refreshing).

5.4 The « Change_dimension() » Function

The « Change_dimension » evolution function is used to make modifications to dimensions in the GAMM model. This evolution can be applied in different scenarios, such as changing a description attribute or a hierarchical link. During this evolution, several cases are considered based on the type of modification:

1. « Change_dimension($\{D, A\}$) » If a description attribute changes in a dimension, the evolution function creates a new node representing this new attribute and links it to the appropriate dimension node. Relationships between dimension nodes are updated by applying a temporal labeling of interval type FD and TD to determine the

validity period of these relationships. FD takes the start date of the relation from the data sources, while TD takes the date 31/12/9999. Additionally, the start date of this new relationship will be assigned to the TD parameter of the old link to determine the end of its validity, defining the validity period of each of these relationships.

2. « Change_dimension($\{D, L\}$) » If a hierarchical link changes between a dimension node and a level node in the hierarchy, the evolution function creates a new link to the appropriate level in the hierarchy. Similar to the previous case, a temporal labeling of interval type FD and TD is applied to this new relationship. FD takes the start date of the relation from the data sources, while TD takes the date 31/12/9999. Additionally, the start date of this new relationship will be assigned to the TD parameter of the old hierarchical link to determine the end of its validity, defining the validity period of each of these relationships.
3. « Change_dimension($\{L, A\}$) » Similar to the case of changing an attribute for a dimension node; if a description attribute changes for a hierarchy level, the evolution function creates a new node representing this new attribute and links it to the appropriate dimension node. Relationships between dimension nodes are updated by applying a temporal labeling of interval type FD and TD to determine the validity period of these relationships. FD takes the start date of the relation from the data sources, while TD takes the date 31/12/9999. Additionally, the start date of this new relationship will be assigned to the TD parameter of the old link to determine the end of its validity, defining the validity period of each of these relationships.
4. « Change_dimension($\{L, L\}$) » Similar to the case of changing a hierarchical link between a dimension node and a level node in the hierarchy; if a hierarchical link changes between two level nodes in the hierarchy, the evolution function creates a new link to the appropriate level in the hierarchy. A temporal labeling of interval type FD and TD is applied to this new relationship. FD takes the start date of the relation from the data sources, while TD takes the date 31/12/9999. Additionally, the start date of this new relationship will be assigned to the TD parameter of the old hierarchical link to determine the end of its validity, defining the validity period of each of these relationships.

5.5 The « Change_fact() » Function

The « Change_fact » evolution function is used to make modifications to fact nodes in the GAMM model in case of errors detected in the data sources. This evolution corrects incorrect or inconsistent values of existing fact nodes. During this evolution, when correcting errors previously loaded from data sources, the punctual timestamp of type VT and TT for the nodes in question is maintained. This means that the initial timestamps of fact nodes remain unchanged even after modifying the values. These modifications can be based on additional data checks, manual corrections, or automatic updates from reliable data sources. By maintaining the punctual timestamp of fact nodes, the GAMM model retains the initial temporal information about these nodes.

The data evolution functions of the GAMM model are designed to ensure a smooth evolution of data instances. The temporal labeling applied to nodes and relationships manages the temporality of the data, ensuring a coherent and historical representation in the data warehouse. This approach enables precise and consistent OLAP queries and analyses, providing a solid foundation for decision-making. We will reuse our motivation example to illustrate the data evolutions described above (Sect. 2).

5.6 Example 1: Evolution of a Hierarchy Level Instance in the CUSTOMER Dimension

Let's analyze the instances of the CUSTOMER dimension presented in Table 1. This table indicates that the customer MARY SAVELEY is associated with two different cities during two distinct time periods. Specifically, the customer resided in PARIS from 01/01/2020 and moved to LYON on 01/08/2020. This change is reflected in the evolution of the hierarchical link (CUSTOMER, CITY), which can be implemented in the GAMM model through the data evolution function *Change_dimension(*CUSTOMER,CITY*)*. We represent this evolution of CUSTOMER dimension instances using the GAMM model formalism, as illustrated in Fig. 4. The FD and TD parameters are utilized to identify the lifespans of hierarchical relations as well as relations with attributes.

The schema shows that the relationship between the customer MARY SAVELEY, with identifier CUST001, and the level PARIS, with identifier CITY001, is valid during the time interval from 01/01/2020 to 01/08/2020, while the relationship between the same customer and the level LYON, with identifier CITY002, is valid from 01/08/2020 and remains valid.

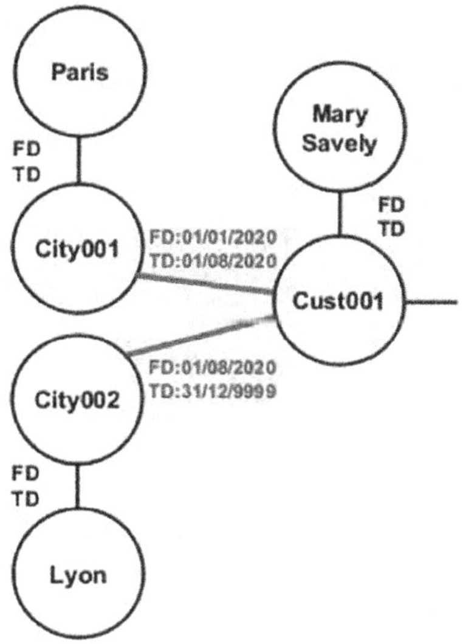

Fig. 4. Example of Changing an Instance in the CUSTOMER Dimension.

The schema illustrates that the relationship between the customer MARY SAVELEY with the identifier CUST001 and the level PARIS with the identifier CITY001 is valid during the time interval from 01/01/2020 to 01/08/2020. Simultaneously, the relationship between the same customer and the level LYON with the identifier CITY002 is valid during the time interval starting from 01/08/2020 and remains ongoing.

5.7 Example 2: Modification of Hierarchy Level in an Instance of the PRODUCT Dimension

We reuse the instances of the PRODUCT dimension presented in Table 2. This table indicates that the product MOZZARELLA is associated with two different categories during two distinct periods. Indeed, the product was initially classified in the FRESH category since 01/01/2020, but it was reclassified into the DAIRY category starting from 01/04/2020.

This change results in an evolution at the level of the hierarchical relationship (PRODUCT, CATEGORY), which can be achieved in the GAMM model through the data function *Change_dimension(*PRODUCT,CATEGORY*)*. We represent this evolution of instances in the PRODUCT dimension using the formalism of the GAMM model, as illustrated in Fig. 5.

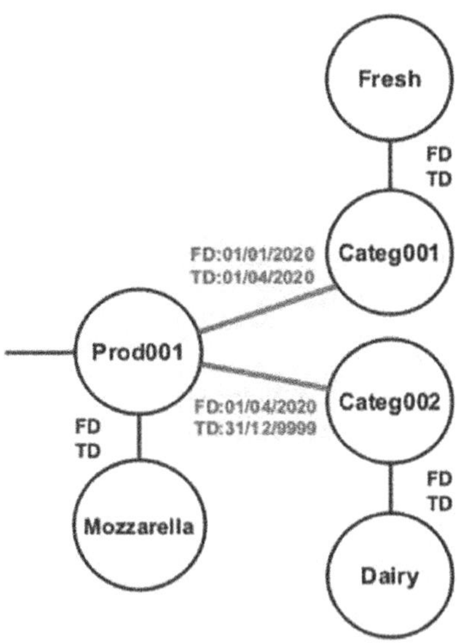

Fig. 5. Example of modification in an instance of the PRODUCT dimension.

This schema illustrates that the hierarchical relationship between the product MOZZARELLA, identified by PROD001, and the level FRESH, identified by CATEG001, is valid during the time period from 01/01/2020 to 01/04/2020. Simultaneously, the relationship between the same product and the level DAIRY, identified by CATEG002, is valid from 01/04/2020 until today.

6 Temporal Queries in GAMM

We represent the instances of our running example, as shown in Tables 1, 2, 3, and 4, using our temporal approach based on a graph, as illustrated in Fig. 6. The FD and TD parameters were employed to identify the lifespans of relationships between dimensions and levels. The labels of nodes and relationship names are excluded from the graph to maintain clarity in the schema.

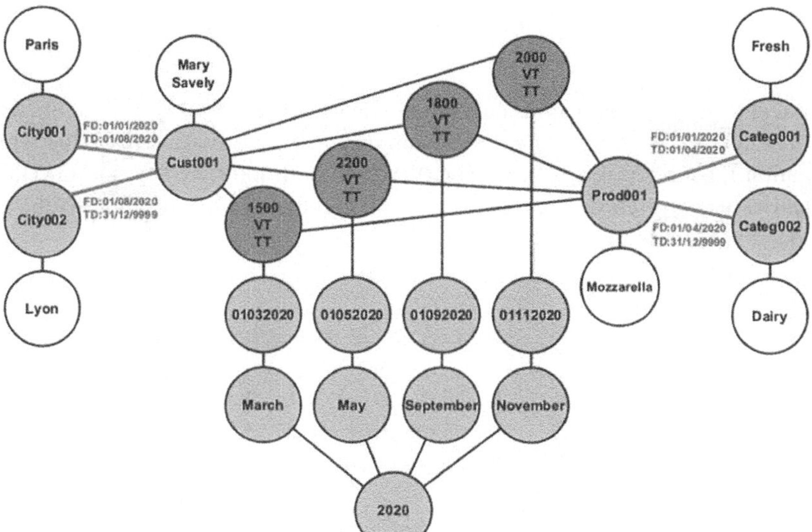

Fig. 6. Example of a data instance from the GTDW.

The schema illustrates that the relationship between the customer *Mary Saveley*, identified by *Cust001*, and the level *Paris*, identified by *City001*, is valid during the time interval from 01/01/2020 to 01/08/2020. Simultaneously, the relationship between this same customer and the level *Lyon*, identified by *City002*, is valid during the time interval from 01/08/2020 to the present. Similarly, the relationship between the product *Mozzarella*, identified by *Prod001*, and the level *Fresh*, identified by *Categ001*, is valid during the time interval from 01/01/2020 to 01/04/2020. Meanwhile, the relationship between this same product and the level *Dairy*, identified by *Categ002*, is valid during the time interval from 01/04/2020 to the present.

The use of a graph database simplifies the implementation of this temporality in relationships between instances, as links also carry information like nodes. In our case, the links between CUSTOMER nodes and CITY nodes contain the attributes FD and TD, enabling the determination of relationship lifespans. This is similarly true for links between PRODUCT nodes and CATEGORY nodes.

Figure 7 depicts the lifespans of relationships involving levels CITY and CATEGORY, with intersections creating three time intervals denoted by T_1, T_2, and T_3. These

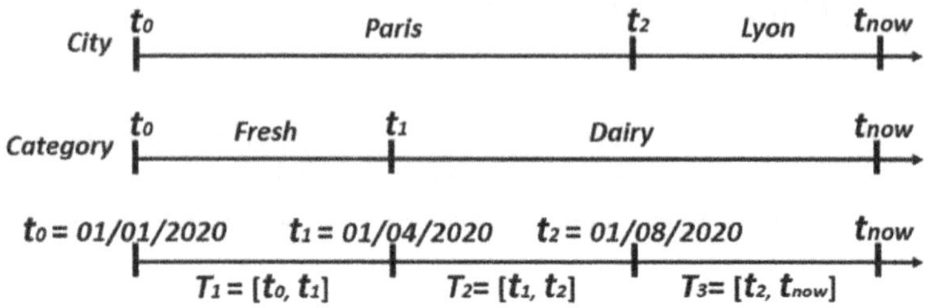

Fig. 7. Timeline of the evolution of levels CITY and CATEGORY.

temporal changes in relationships necessitate the use of temporal queries to obtain consistent results. Below, we present three examples of illustrative queries using temporal aggregation for the levels CITY, CATEGORY, and both jointly CITY and CATEGORY. The queries are written in the Cypher query language (CQL), specific to the Neo4j graph database (Table 6).

Query 1:

```
MATCH (cn:city_name)<-[]-(:city)<-[r:customer_city]-(c:c⌋
↪  ustomer)<-[]-(s:sales)
WHERE r.From_Date <= s.valid_date < r.To_Date
RETURN c.customer_id, cn.city_name, SUM(s.sales_amount)
```

Table 6. Results of Query 1.

Customer_ID	City_Name	Sales_Amount
Cust001	Paris	3700
Cust001	Lyon	3800

Query 1 retrieves the sales amounts for each city for the customer *Cust001*. Results are obtained for each city the customer lived in. Note that in CQL, the clause [RETURN $value_1$,.., $value_n$, AGGREGATE_FUNCTION(ATTRIBUTE)] allows grouping the aggregation by $value_1$... $value_n$ (Table 7).

Query 2:

```
MATCH (cn:category_name)<-[]-(:category)<-[r:product_cat⌋
↪  egory]-(p:product)<-[]-(s:sales)
WHERE r.From_Date <= s.valid_date < r.To_Date
RETURN p.product_id, cn.category_name,
↪  SUM(s.sales_amount)
```

Query 2 retrieves the sales amounts for each category for the product *Prod001*. Results are obtained for each category the product was assigned to (Table 8).

Table 7. Results of Query 2.

Product_ID	Category_Name	Sales_Amount
Prod001	Fresh	1500
Prod001	Dairy	6000

Query 3:

```
MATCH (ctn:city_name)<-[]-(:city)<-[r1:customer_city]-(c
↪   :customer)<-[]-(s:sales)-[]->(p:product)-[r2:product
↪   _category]->(:category)-[]->(cgn:category_name)
WHERE r1.From_Date <= s.valid_date < r1.To_Date
AND r2.From_Date <= s.valid_date < r2.To_Date
RETURN ctn.city_name, cgn.category_name,
↪   SUM(s.sales_amount)
```

Table 8. Results of Query 3.

City_Name	Category_Name	Sales_Amount
Paris	Fresh	1500
Paris	Dairy	2200
Lyon	Dairy	3800

Query 3 retrieves the sales amounts for different cities and categories. Results are obtained for each city the customer lived in and for each category the product was assigned to.

By the use of temporal parameters `From_Date` and `To_Date`, accurate results are obtained regardless of changes made in dimension instances. The same principle applies to changes in dimension attribute values due to the separation of business concepts from their descriptors (attributes). The temporal parameters FD and TD are also implemented in the relationships between these attributes. The graph database's ability to use information about relationships between entities provides advantages when formulating temporal queries.

7 Case Studies for Validation and Runtime Performance

For validation purposes, we conducted two case studies based on the Star Schema Benchmark (SSB). The first study aimed at functionally validating our approach, while the second focused on runtime tests for a graph Data Warehouse. In the functional validation study, we instantiated the SSB data in Neo4j, applying temporal changes in aggregation relationships of hierarchical levels and attributes. We generated hierarchical levels from the attributes, as depicted in Fig. 8. The SSB queries were then executed on this temporal DW.

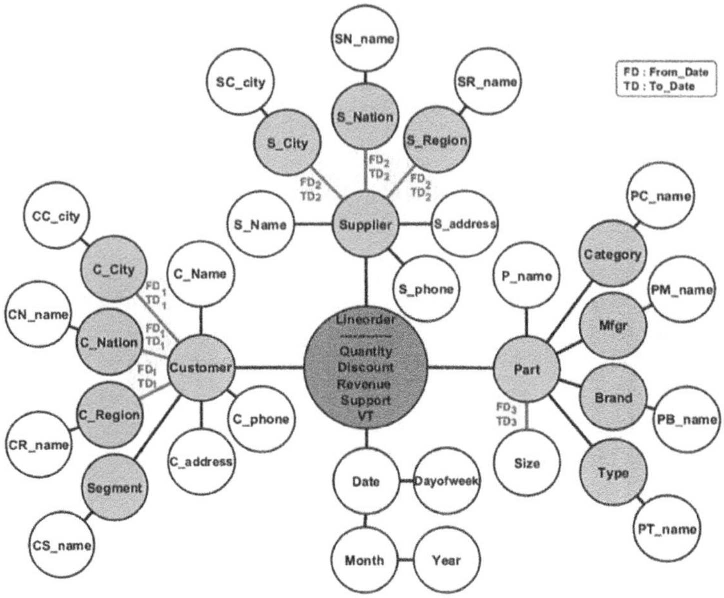

Fig. 8. Logical schema for the SSB graph temporal DW.

Temporal changes were introduced for the dimensions CUSTOMER, SUPPLIER, and PART as follows:

1. For the dimension CUSTOMER, assignments of one hundred randomly selected customers were changed on the hierarchical levels C_CITY, C_NATION, and C_REGION for two time intervals: T_1 and T_1'.
2. For the dimension SUPPLIER, assignments of one hundred randomly selected suppliers were changed on the hierarchical levels C_CITY, C_NATION, and C_REGION for two time intervals: T_2 and T_2'.
3. For the dimension PART, the SIZE attributes of one hundred randomly selected products were changed for two time intervals: T_3 and T_3'.

We used Query 3.3 from the SSB queries as an illustrative example.

```
OPTIONAL MATCH (c:cc_city) <-[:c_c_name]-(:c_city)<-[r1:customer_city]-(:customer)<-[:order_customer]-(l:lineorder)-[:order_date]->(d:date),
(s:sc_city)<-[:s_c_name]-(:s_city)<-[r2:supplier_city]-(s:supplier)
<-[:order_supplier]-(l)
WHERE r1.From_Date<=l.valid_date<r1.To_Date
AND r2.From_Date<=l.valid_date<r2.To_Date
AND 1992<= d.D_YEAR<=1997
AND (c.c_city="united ki1" OR c.c_city="united ki5")
```

```
AND (s.s_city="united ki1" OR s.s_city = "united ki5")
RETURN c.c_city, s.s_city, d.d_year, SUM(l.lo_revenue)
↪   AS revenue
ORDER BY d.d_year ASC, revenue DESC
```

The above query requires aggregating the CITY levels of the dimensions CUSTOMER and SUPPLIER. The query has been conditioned according to the FD and TD parameters of these relationships to obtain consistent results. We successfully applied all 13 SSB queries in a temporal format, verifying the accuracy of the results. The dataset, GTDW creation commands, and all queries written in Cypher are available on GitHub.

For the runtime performance study, we compared the execution times of the 13 queries on both relational and graph approaches. The SSB schema was generated using our graph approach, and the execution times were compared using Neo4j 4.4.5 for the graph approach and Oracle 11g for the relational approach. The average execution times for ten runs of each query using both approaches are displayed in Fig. 9. The graph approach outperformed the relational approach for several queries, while the relational approach performed better for others. These differences depend on filter factors (FFs), the number of dimensions, and the number of edges to be covered.

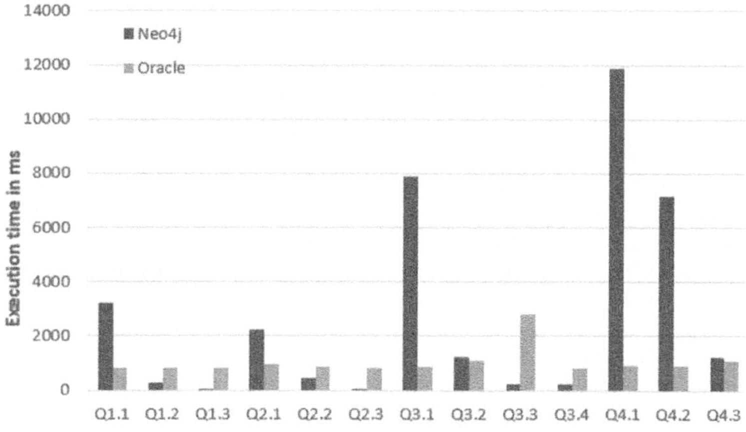

Fig. 9. Response times for SSB queries using the graph and relational approaches.

We observed from the results that the graph approach outperformed the relational approach for queries Q1.2, Q1.3, Q2.2, Q2.3, Q3.3, and Q3.4 (showing improvements from 2× to 11×). The performance of the two approaches was closely matched, with the relational approach being slightly faster for queries Q3.2 and Q4.3. The relational approach demonstrated better performance for queries Q1.1, Q2.1, Q3.1, Q4.1, and Q4.2 (with improvements ranging from 2× to 8×). The variations in execution times between the graph approach and the relational approach are influenced by filter factors, the number of dimensions, and the number of edges to be covered. Neo4j exhibits high efficiency when FF is low, but its efficiency may decrease as FF and the number of dimensions increase. This is because navigating numerous relations becomes necessary for performing aggregation functions. In contrast, Oracle demonstrates a more consistent performance in terms of query execution time. It's important to note that within the context of temporal approaches, a comprehensive performance study should be conducted, encompassing the entire integration and storage process.

8 Conclusion

In our previous work [6], we introduced a temporal multidimensional model for multiversion data warehouses based on a graph formalism. In this paper, we complement our approach by proposing temporal data evolution functions and elucidating the principle of temporal management in GAMM. Temporal labeling was applied to entities and their relationships, enabling optimal data instance management and preserving the evolutionary history, particularly for changes in dimensions. This approach offers the flexibility of schema evolution provided by the multi-versioning concept in data warehouses. Our approach was validated through functional testing and runtime performance experiments in two use cases. Our future work includes a comprehensive performance study of the entire integration process, further exploration of OLAP queries and graph cubes, and an investigation into advanced analyses in graph models for online analysis to generate explanatory and predictive models.

References

1. Ahmed, W., Zimányi, E., Vaisman, A.A., Wrembel, R.: A temporal multidimensional model and OLAP operators. Int. J. Data Warehous. Min. **16**(4), 112–143 (2020). https://doi.org/10.4018/IJDWM.2020100107
2. Ahmed, W., Zimányi, E., Wrembel, R.: Temporal data warehouses: logical models and querying. In: Zimányi, E., Vansummeren, S., Calders, T. (eds.) Actes des 11es journées francophones sur les Entrepôts de Données et l'Analyse en Ligne. RNTI, vol. B-11, pp. 33–48 (2015)
3. Akoka, J., Comyn-Wattiau, I., du Mouza, C., Prat, N.: Mapping multidimensional schemas to property graph models. In: Reinhartz-Berger, I., Sadiq, S. (eds.) ER 2021. LNCS, vol. 13012, pp. 3–14. Springer, Cham (2021). https://doi.org/10.1007/978-3-030-88358-4_1
4. Benhissen, R., Bentayeb, F., Boussaid, O.: GAMM: un modèle multidimensionnel agile à base de graphes pour des entrepôts multi-versions. Revue des Nouvelles Technologies de l'Information. Business Intelligence & Big Data, RNTI-B-18, pp. 29–46 (2022)
5. Benhissen, R., Bentayeb, F., Boussaid, O.: GAMM: graph-based agile multidimensional model. In: Gallinucci, E., Golab, L. (eds.) Proceedings of the 25th International Workshop on Design, Optimization, Languages and Analytical Processing of Big Data (DOLAP) co-located with (EDBT/ICDT), Ioannina, Greece, 28 March 2023, pp. 23–32. CEUR Workshop Proceedings, CEUR-WS.org (2023)
6. Benhissen, R., Bentayeb, F., Boussaïd, O.: Temporal multidimensional model for evolving graph-based data warehouses. In: Gusikhin, O., Hammoudi, S., Cuzzocrea, A. (eds.) Proceedings of the 12th International Conference on Data Science, Technology and Applications, DATA 2023, Rome, Italy, 11–13 July 2023, pp. 40–51. SCITEPRESS (2023). https://doi.org/10.5220/0012080400003541

7. Bliujute, R., Saltenis, S., Slivinskas, G., Jensen, C.S.: Systematic change management in dimensional data warehousing. In: Proceedings of the Third International Baltic Workshop on Data Bases and Information Systems, Riga, Latvia. Citeseer (1998)
8. Böhlen, M.H., Dignös, A., Gamper, J., Jensen, C.S.: Temporal data management - an overview. In: Zimányi, E. (ed.) eBISS 2017. LNBIP, vol. 324, pp. 51–83. Springer, Cham (2017). https://doi.org/10.1007/978-3-319-96655-7_3
9. Campos, A., Mozzino, J., Vaisman, A.: Towards temporal graph databases. arXiv preprint. arXiv:1604.08568 (2016)
10. Debrouvier, A., Parodi, E., Perazzo, M., Soliani, V., Vaisman, A.A.: A model and query language for temporal graph databases. VLDB J. **30**, 825–858 (2021)
11. Faisal, S., Sarwar, M.: Handling slowly changing dimensions in data warehouses. J. Syst. Softw. **94**, 151–160 (2014)
12. Faisal, S., Sarwar, M., Shahzad, K., Sarwar, S., Jaffry, S.W., Yousaf, M.M.: Temporal and evolving data warehouse design. Sci. Program. **2017**, 7392349:1–7392349:18 (2017). https://doi.org/10.1155/2017/7392349
13. Garani, G., Adam, G.K., Ventzas, D.: Temporal data warehouse logical modelling. Int. J. Data Min. Model. Manag. **8**(2), 144–159 (2016)
14. Golfarelli, M., Maio, D., Rizzi, S.: Conceptual design of data warehouses from E/R schema. In: Thirty-First Annual Hawaii International Conference on System Sciences. IEEE Computer Society (1998)
15. Golfarelli, M., Rizzi, S.: Managing late measurements in data warehouses. Int. J. Data Warehous. Min. **3**(4), 51–67 (2007)
16. Golfarelli, M., Rizzi, S.: Temporal data warehousing: approaches and techniques. In: Integrations of Data Warehousing, Data Mining and Database Technologies – Innovative Approaches. Information Science Reference (2011)
17. Golfarelli, M., Rizzi, S.: From star schemas to big data: 20+ years of data warehouse research. In: Flesca, S., Greco, S., Masciari, E., Saccà, D. (eds.) A Comprehensive Guide Through the Italian Database Research Over the Last 25 Years. SBD, vol. 31, pp. 93–107. Springer, Cham (2018). https://doi.org/10.1007/978-3-319-61893-7_6
18. Inmon, W.H.: Building the Data Warehouse. Wiley, USA (1992)
19. Jensen, C.S., Snodgrass, R.T.: Transaction time. In: Liu, L., Özsu, M.T. (eds.) Encyclopedia of Database Systems, 2nd edn., pp. 4200–4201. Springer, New York (2018). https://doi.org/10.1007/978-1-4614-8265-9_1064
20. Jensen, C.S., Snodgrass, R.T.: Valid time. In: Liu, L., Özsu, M.T. (eds.) Encyclopedia of Database Systems, 2nd edn., pp. 4359–4360. Springer, Nwe York (2018). https://doi.org/10.1007/978-1-4614-8265-9_1066
21. Kimball, R.: The Data Warehouse Toolkit: Practical Techniques for Building Dimensional Data Warehouses. Wiley, USA (1996)
22. Kimball, R., Ross, M.: The Data Warehouse Toolkit: The Definitive Guide to Dimensional Modeling, 3rd edn. Wiley, Indianapolis (2013)
23. Mendelzon, A., Vaisman, A.: Temporal queries in OLAP. In: Proceedings of the 26th International Conference on Very Large Data Bases, VLDB 2000 (2000)
24. Phungtua-Eng, T., Chittayasothorn, S.: Slowly changing dimension handling in data warehouses using temporal database features. In: Nguyen, N.T., Gaol, F.L., Hong, T.-P., Trawiński, B. (eds.) ACIIDS 2019. LNCS (LNAI), vol. 11431, pp. 675–687. Springer, Cham (2019). https://doi.org/10.1007/978-3-030-14799-0_58
25. Saroha, K., Gosain, A.: Bi-temporal schema versioning in bi-temporal data warehouse. CSI Trans. ICT **3**, 135–142 (2015)

Data Science for the Promotion of Sustainability in Smart Water Distribution Systems

Sarah Di Grande[1(✉)], Mariaelena Berlotti[1], Salvatore Cavalieri[1], and Roberto Gueli[2]

[1] University of Catania, 95125 Catania, CT, Italy
{sarah.digrande,mariaelena.berlotti}@phd.unict.it, salvatore.cavalieri@unict.it
[2] EHT, Viale Africa 31, 95129 Catania, Italy
roberto.gueli@eht.eu

Abstract. In today's global landscape, the confluence of diminishing water reservoirs and the unprecedented surge in energy expenses represents a critical juncture in the discourse of sustainable resource management. In this context, Water Distribution Systems are changing their approach for the distribution of water to consumers. The sustainable management of water aligns with the Watergy concept, emphasizing the fulfillment of user demand while optimizing the utilization of water and energy resources. Leveraging modern technologies such as Artificial Intelligence, Internet of Things, Big Data Analytics, the proactive forecasting of water and energy demand stands as a fundamental strategy for achieving this objective. The significance of water demand forecasting extends beyond anticipating resource allocation for water distribution companies. It serves as a proactive measure, facilitating the early detection of consumption irregularities and playing a crucial role in scheduling pumps. Conversely, energy consumption forecasting serves diverse functions, including energy optimization, anomaly identification in consumption patterns, and effective energy load planning.

This study aims to pioneer short-term forecasting models for water demand and energy consumption within the water distribution sector, employing innovative machine learning-based methodologies. The utilization of global forecasting models, the N-Beats machine learning algorithm, and transfer learning techniques demonstrates exceptional promise in creating robust predictive models for this domain, exhibiting notable performance.

Keywords: Water 4.0 · Energy · Sustainability · Forecasting · Machine Learning

1 Introduction

1.1 Watergy Efficiency

The scarcity of water and rising energy costs are compelling Water Distribution Systems (WDSs) to reconsider their approach to supplying water to consumers.

The looming threat of water scarcity presents a formidable risk to humanity's well-being. Projections suggest that by 2025, approximately 1.8 billion individuals could

O. Gusikhin et al. (Eds.): DATA 2023, CCIS 2105, pp. 50–72, 2024.
https://doi.org/10.1007/978-3-031-68919-2_3

confront acute water shortages, while nearly two-thirds of the global population might grapple with water stress [1]. This diminishing water availability stems from a confluence of factors, including climate change, burgeoning population growth, escalating rates of urbanization, and the expansion of industrial activities [2–5]. These elements have substantially elevated water consumption, leading to a depletion of available water resources [6, 7]. Importantly, the surge in water consumption has occurred without a parallel increase in water resources. Moreover, should the present trajectory persist, water demand is projected to surpass supply by 40% by the year 2030 [8].

Energy is intrinsically dependent on water because, without energy, water cannot be supplied to a system. Approximately 7%–8% of the total energy generated worldwide is utilized for the production and distribution of drinking water [9]. The majority of this energy is derived from fossil fuels, leading to environmental pollution. Given economic and population growth, energy consumption is poised to increase [10]. Driven by goals of sustainability, cost savings, and adherence to environmental regulations, companies today are genuinely motivated to reduce energy consumption.

For these reasons, WDSs operator are tasked with delivering water to users that meets specific quality, pressure, and quantity standards while striving to minimize both water and energy usage to prevent wastage [8, 11]. This objective is intricately tied to the notion of Watergy efficiency, aiming to fulfill user demand while utilizing water and energy resources in the most efficient manner possible [12]. This balance can be attained through operational actions that necessitate the application of forecasting models. These optimized operations have the potential to yield savings ranging from 25% to 30% by reducing electricity and treatment inputs [4].

1.2 Artificial Intelligence in Water Distribution Systems

To reach these goals, an industrial revolution of the water sector was needed [8]. The Water 4.0 revolution has introduced different features, among which automation, an increased integration of sensors, Internet of Things, Big Data analysis, and Artificial Intelligence. The most famous applications of artificial intelligence in the management of WDSs are anomaly detection and water demand forecasting [8].

Predicting water demand is crucial for detecting inefficient practices or system leaks, resulting not only in increased water usage but also in higher energy consumption [13–15]. In the short term, predictive models help in the optimal performance of wells and reservoirs, for the decision-making about the allocation of resources, and an optimal water pressure management [16]. Moreover, forecasting water demand helps prevent energy wastage by enabling the optimal management of pump operations, also called pump scheduling [16–18]. Concerning energy usage, a water distribution system incurs substantial energy expenses across all its functions, encompassing water extraction, treatment, and distribution [19]. However, pumping systems stand out as the primary contributors to this consumption [20]. To comprehend how water demand can decrease energy consumption, Fig. 1 can be utilized. Figure 1 illustrates the distribution of water to end-users within a simplified water distribution network. In this network, individual zones are served by a pump system consisting of three parallel pumps that collaborate to distribute water sourced from a reservoir. Following each pump system, a single pipe facilitates the water's flow to a specific network zone. The parallel pump systems present opportunities for energy conservation by allowing decisions on when to activate or deactivate pumps based on both water demand and energy consumption [17]. In simpler terms, not all pumps are required to operate simultaneously; for instance, during nighttime when water demand typically decreases compared to regular working hours.

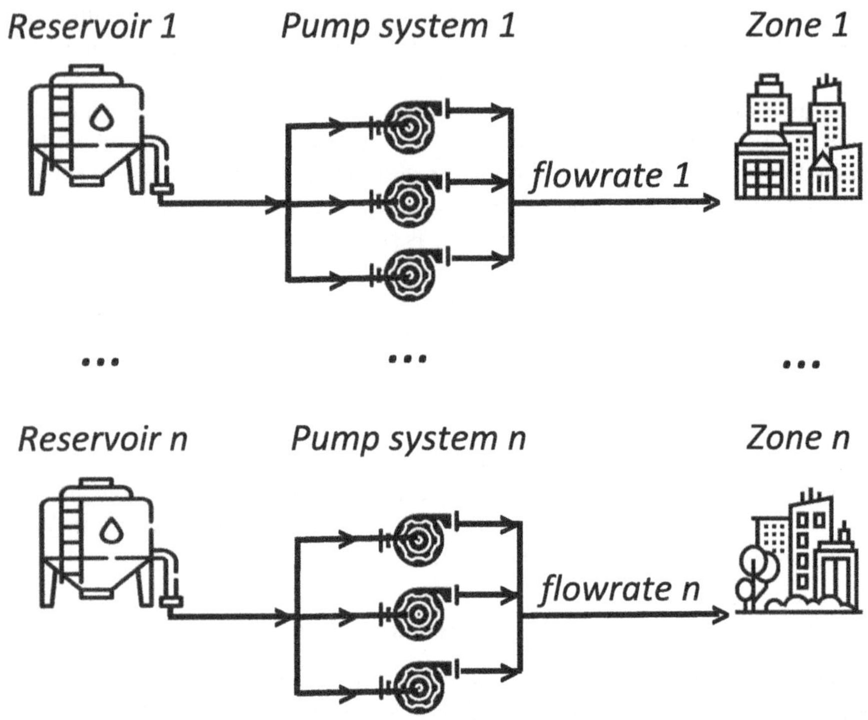

Fig. 1. A simple representation of a Water Distribution Network in which the water is distributed to different zones of a city through a system of parallel pumps extracting water from a reservoir [21].

Therefore, predicting the water demand for each network zone assumes a critical role in scheduling pump operations. Precision in forecasting aids in determining the quantity and selection of pumps to activate at specific times of the day based on water demand. In a water distribution network, the importance lies not only in forecasting water demand but also in forecasting energy consumption. Anticipating the energy consumption of pump systems serves multiple purposes, including energy optimization, identification of abnormal consumption patterns, and formulation of energy load plans [22, 23]. Particularly, planning the energy load to be managed is crucial for estimating anticipated costs and assessing the system's capacity to meet the required power demands.

1.3 State of the Art

From the existing literature, papers that addressed water demand and energy consumption forecasting in water distribution systems were specifically chosen.

In [4, 16] the authors provide an extensive examination of the utilized methodologies and significant forthcoming challenges. Specifically, the top three commonly employed techniques are as follows: traditional time series algorithms (such as autoregressive integrated moving averages and exponential smoothing), various artificial neural network variants (including long short-term memory, radial basis function ANN, and gated recurrent units), and support vector machines.

Time series models, unlike machine learning models, often fall short of achieving highly accurate forecasts due to their limitations in comprehending intricate nonlinear patterns within demand forecasting. Nevertheless, these models remain widely utilized owing to their simplicity and ease of interpretation [16]. Among recent studies, AutoRegressive Integrated Moving Averages (ARIMA) and Exponential Smoothing (ES) emerge as the predominant time series models. For instance, in the study [24] two forecasting methods were employed for daily urban water consumption forecasting, one of which included ARIMA. Similarly, [25] explored diverse methodologies, incorporating ARIMA and ES, to achieve mid-term forecasts. Moreover, [26] utilized ARIMA and ES approaches for forecasting monthly urban water demand.

Machine learning methods, such as ANN, are widely applied for water demand forecasting in complicated water systems because of their ability to deal with nonlinearities. In [27] statistical methods and conventional ANNs were compared with deep learning methods, proving that deep learning models give more accurate short-term water demand forecasts. In [28] hourly water demand forecasting was implemented through the Gated Recurrent Unit (GRU) method, demonstrating the feasibility of using this architecture. The GRU architecture to make an hourly water demand forecasting was also used in [29], demonstrating the superiority of this model compared to Support Vector Machines.

The SVM model with a polynomial kernel function was also used in [30], to predict monthly water demand, highlighting that a small number of studies have used this technique in predicting water demand [31, 32]. Furthermore, in [33] the authors used the SVR with a parallel global optimization tuning of hyperparameters to improve the accuracy of short-term forecasts.

Another class of methods used for forecasting is the hybrid approach. These methods involve combining multiple forecasting techniques or models, such as artificial neural networks with traditional time series models or integrating machine learning and deep learning algorithms, to enhance the accuracy and reliability of predictions [16, 34]. By leveraging the strengths of different approaches, they compensate for each other's weaknesses, ultimately providing more robust forecasts. In studies [35, 36], hybrid methods were employed based on a traditional ARIMA approach combined with ANNs to predict urban water demand. The ARIMA model analyzed the linear part of the problem, while the ANN modeled residuals. The proposed hybrid model demonstrated superior predictive capability compared to using ANN or ARIMA models separately in forecasting water demand.

Forecasting the energy consumption of pumps in a water distribution system remains a relatively unexplored research area. In a study [22], the authors utilized four different algorithms—multiple linear regression, random forest, deep neural network, and support vector regression—to predict the energy consumption of the entire water system and its individual subsystems: conveyance, treatment, and distribution. The authors highlighted the scarcity of similar literature focusing on this specific aspect. While there is a dearth of studies focusing on the distribution phase, other studies have concentrated on forecasting energy consumption in wastewater treatment [37–39]. In [10], a detailed review was conducted encompassing various methods and application areas for the general topic of energy consumption forecasting. Similar to water demand forecasting, in the field of energy, the most commonly utilized methods are traditional time series models and AI-based models, notably Artificial Neural Networks (ANN) and Support Vector Regression (SVR) [40–43]. In terms of usage proportion, 77% of the papers employ ANN models, while 23% utilize SVR models; additionally, short-term forecasting is predominantly favored with these models [10]. Even in these studies, hybrid models were used.

Based on the conducted research and insights highlighted in [16], several underexplored areas exist within the existing literature. In typical water distribution systems, the forecasting of water demand time series involves developing individual models for each time series. However, a shift towards global models is proposed, as advocated in [44], which could enable the creation of a single model encompassing all time series. This becomes imperative due to the continuous increase in the number of available series facilitated by advancing technologies, necessitating new tools for effective management. Moreover, global models, capable of learning from multiple time series concurrently, offer the opportunity to employ transfer learning methodologies, as emphasized in [45]. When trained on diverse series, transfer learning enables forecasting on previously unencountered series. Lastly, to the best knowledge of the authors, among the various algorithms employed, one particular algorithm remains unexplored in the context of water demand and pump energy consumption forecasting within an entire water distribution system: N-Beats [46]. Developed explicitly for time series data, this algorithm presents an untapped potential for this domain.

1.4 Paper Aim

In summary, the present paper is focused on the water demand and the pumps' energy consumption forecasting, with the implementation of a multi-step ahead forecasting where 24 h of consumption are predicted. The paper aims to do short-term forecasting through global machine learning models, the N-Beats machine learning algorithm, and transfer learning approaches.

This paper constitutes an extended version of the work previously published in [21]. The authors have expanded upon the earlier research by presenting more comparisons with the literature and more detailed results compared to those previously showcased. Additionally, they have conducted more experiments testing the global N-Beats model for one-step forecasting with a Walk Forward Validation technique, an aspect that was previously limited to multi-step forecasting.

The proposal outlined in the paper emerged from a research project funded by the "Italian Ministry of Enterprises and Made in Italy" (https://www.mimit.gov.it/en/). Comprehensive details about this research are provided in the Acknowledgment section. Among the project's collaborators, EHT S.C.p.A. stands out as an enterprise deeply engaged in the digital transformation of water distribution systems. The objectives outlined in this research were defined by EHT in conjunction with the project sponsors, all of whom are extensively involved in the same industry. Additionally, the participating enterprises evaluated the attained results to gauge their practical significance. This evaluation proved successful, as the proposal's impact on the effective management of water distribution systems was deemed substantial.

The paper's structure is as follows: Sect. 2 delineates the authors' proposed approach to address the earlier mentioned forecasting issues, outlining the simulation of the dataset, and preprocessing steps preceding the machine learning algorithm; Sect. 3 describes the suggested forecasting models, validation process, and the performance metrics employed for model evaluation; Sects. 4 and 5 present the achieved results and conclude along with prospects for future work, respectively.

2 Data

Within this section, the authors delineate the utilized tools for data simulation and the stages of dataset preprocessing.

2.1 Water Distribution Simulation

Data are recognized to be crucial in the application of a machine learning strategy. In the context of the problem highlighted within this paper, data concerning the pipe flow rates and pump energy usage of WDS is essential. Frequently, data associated with water distribution systems is either inadequately available or lacks quality [47]. Therefore, given that the primary objective of the paper was to assess the viability of the proposed approach, synthetic data were generated to enable the execution of the machine learning-based solution.

The generation of data was conducted using the Water Network Tool for Resilience (WNTR), an EPANET-based Python package engineered for modeling water distribution networks (WDNs), version 0.5.0 [48]. Through the WNTR simulator, an.inp file is utilized to input network attributes (such as pipes, pumps, valves, junctions, tanks, reservoirs, water demand patterns, and pump curves), producing various time series presenting simulation outcomes (including pipe flow rates, node pressures, and pump energy consumption).

The simulation utilized a simplified version of the water distribution network in Milan, Italy, comprising 12,354 nodes, 17,548 links, 128 patterns, and 95 unique pump curves (one per pump). Each node, representing a user, was allocated a specific water demand pattern. These patterns were derived from minute-by-minute water consumption data collected on specific dates. To elaborate, these consumption records were consolidated into a dataset where individual columns denoted distinct zones, and each row contained the water usage for a particular minute of the day. To derive coefficients for the overall demand pattern, a column was introduced, capturing the total consumption per minute. Subsequently, the coefficients per minute were computed by dividing each entry in the total consumption column by the average value of that column. These coefficients were then aggregated using the mean operator to generate observations at one-hour intervals, facilitating subsequent hourly forecasting steps.

Derived from the genuine demand pattern, four additional patterns were generated by introducing random noise to each value of the original pattern. To create akin yet distinct patterns, a quantity that either increased or decreased the initial value of the authentic pattern was added. Calculating this quantity involved generating random noise within the range of -1.5 to 1.5. This was achieved by multiplying a random value between 0 and 1 by 3 and then subtracting 1.5. Subsequently, this value was multiplied by 0.05 to produce a relative number within the 5% range of -1.5 to 1.5. This process was iterated four times for each value in the real pattern, resulting in four similar but distinct patterns.

Although this represents the general water demand pattern, each node within the network.inp file has a distinct base demand, rendering the demand pattern unique for every node in the network.

Moreover, the pump's speed was maintained constant as the current version of WNTR does not facilitate variable speed pumps.

The simulation of water distribution employed this network to generate hourly time series data for a span of 5 days (120 observations). It was a demand-driven simulation, signifying that the system's pressure relied on node demands, ensuring that node demands were consistently met [48]. The gathered time series data intended for subsequent analysis include: flow rate series for each pipe (120 rows, 17,548 columns), and energy consumption readings for each pump (120 rows, 95 columns).

Table 1 illustrates the flow rate dataset, and Table 2 displays the energy consumption dataset. Within these tables, 'f' denotes flow rate, 'e' signifies energy, and the row indices correspond to the hours (totaling 120 h due to the inclusion of 5-day hourly data).

Table 1. Simulated flowrates time series example [21].

Rows index	Flow pipe id 1	Flow pipe id 2	…	Flow pipe id 17548
1	$f_{1,1}$	$f_{1,2}$	…	$f_{1,17548}$
2	$f_{2,1}$	$f_{2, 2}$	…	$f_{2,17548}$
…	…	…	…	…
120	$f_{120,1}$	$f_{120,2}$	…	$f_{120,17548}$

Table 2. Simulated energy consumptions time series example [21].

Rows index	Energy pump id 1	Energy pump id 2	…	Energy pump id 95
1	$e_{1,1}$	$e_{1,2}$	…	$f_{1,95}$
2	$e_{2,1}$	$e_{2, 2}$	…	$f_{2,95}$
…	…	…	…	…
120	$e_{120,1}$	$e_{120,2}$	…	$f_{120,95}$

2.2 Data Preprocessing

A feature selection process was conducted on both the flow rates dataset and the dataset concerning pumps' energy consumption, aiming to determine the most optimal input configuration for the machine learning algorithms.

As depicted in Fig. 1, subsequent to each set of pumps, water flows through a pipe to a zone based on the aggregate water demand within that zone. Following this primary pipe, numerous other pipes facilitate water distribution to all end users. Given the specific interest in the aggregate demand within a network area, the pipe indicated in Fig. 1 was singled out for each served zone using its unique identifier. Consequently, the flow rates dataset was reduced from 17,548 to 27 columns, with each column representing one pipe for each network zone. Among these 27 time series, seven exhibited a nearly constant pattern and were therefore excluded from further analysis. This led to a subsequent reduction from 27 to 20 columns by excluding pipes associated with areas displaying an almost constant daily pattern.

The identified zones within the Milan network include: Anfossi, Armi, Assiano, Baggio, Cantore, Chiusabella, Cimabue, Comasina, Crescenzago, Feltre, Gorla, Italia, Lambro, Novara, Padova, S. Siro, Salemi, Suzzani nord, Suzzani sud, and Vialba. Consequently, the final dataset comprised 20 time series, each one representing the aggregated water demand over a 5-day period for a network zone with hourly intervals.

For the pumps' energy consumption dataset, an initial reduction of features was necessary to retain only the energy consumption of pumps within the previously selected network zones, reducing the count from 95 to 72 pumps. A file associating each pump with its corresponding area was consulted to facilitate this selection process. In each area, multiple pumps collaborated in delivering water to the specific zone, with the number of collaborating pumps varying from a minimum of 2 to a maximum of 5. Additionally, the pump consumption within the same zone was aggregated using the sum operator to prepare the dataset for forecasting the energy consumption of pumps within a network zone.

Similar to the water demand dataset, the final dataset for pumps' energy consumption comprised 20 time series, each one representing the aggregated energy consumption of pumps over a 5-day period for a network zone with hourly intervals.

3 Methodology

In this section, the authors describe the proposed forecasting methodology, and the criteria employed for evaluating the models performance.

3.1 Forecasting Approaches

The paper aims to devise machine learning (ML) models to forecast the aggregated water demand and pumps' energy consumption across 20 distinct water distribution network zones.

To standardize each time series, a Min-Max scaling method was applied, ensuring that the range of each variable fell between 0 and 1. Specifically, using 'max' as the maximum value and 'min' as the minimum value of a variable, the transformation of each observation 'x' was carried out using the formula:

$$\frac{x - \min}{\max - \min} \tag{1}$$

For the preparation of datasets for the machine learning algorithms, each time series was divided into a training set (comprising the initial 4 days, equivalent to 96 h of observations) and a test set (consisting of the final day, encompassing 24 h of observations).

The proposed forecasting approach is delineated in Fig. 3 and Fig. 4. All ML models were executed using Darts (version 0.23.1), a Python library tailored for time series analysis and forecasting [49]. Darts boasts modern ML functionalities with an intuitive API design, offering robustness and ease of use [49]. Additionally, the deep learning forecasting models in Darts are global, exhibiting the potential to be trained with multiple time series, thus enabling forecasting for both trained and unseen series through a transfer learning approach.

Detailed information on the training of global forecasting models and the architecture of the utilized model will be expounded upon in the subsequent paragraph.

Figure 2 and Fig. 3 illustrate the water demand forecasting methodologies applied to each flow rate time series using two distinct approaches. The first method involves crafting a global forecasting model that takes all the flow rate series' training sets as input and provides a 24-h forecast for these series collectively (depicted in Fig. 2). The N-Beats model in Darts was employed for forecasting, with default hyperparameters, except for two: the 'input_chunk_length' set to 72, implying the neural network assesses the past 72 h, and the 'output_chunk_length' set to 24, signifying the neural network generates a 24-point forecast. The second method entails creating a global forecasting model for each flow rate series, employing all flow rate series for training except for the one being forecasted. This technique leverages knowledge acquired from training some series to forecast unseen ones. Figure 3 showcases an instance of this approach applied to forecast a single flow rate series. The latter method was also utilized to construct the pumps' energy consumption forecasting model. As the energy consumption of pumps is closely tied to the flow rate pattern, the WNTR simulator computes it based on the pump flow rate, node head, and the pump's global efficiency (set to 75%). Thus, each pump's energy consumption time series displays a similar pattern to the flow rate time series within the same network zone. Instead of creating a separate model, the first global model trained with all flow rate time series was tested for the energy consumption forecasting of each zone (transfer learning technique), as displayed in Fig. 2.

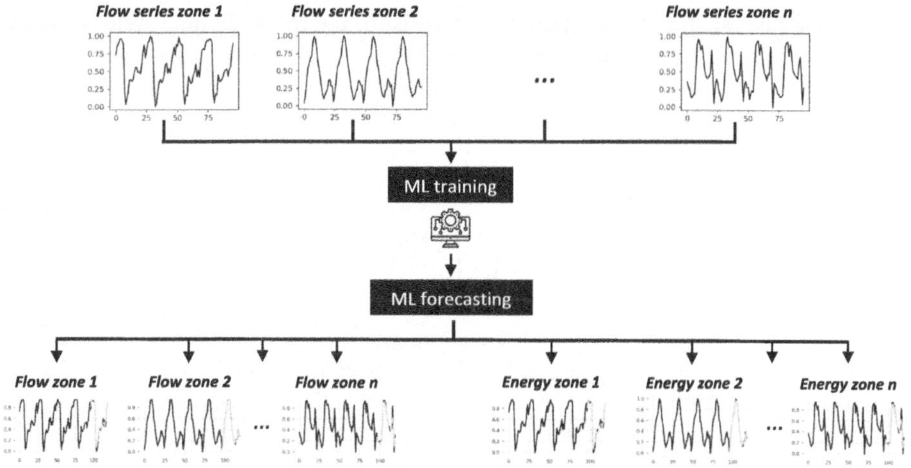

Fig. 2. Water demand and pumps' energy consumption forecasting through a unique ML model trained with flowrates series [21].

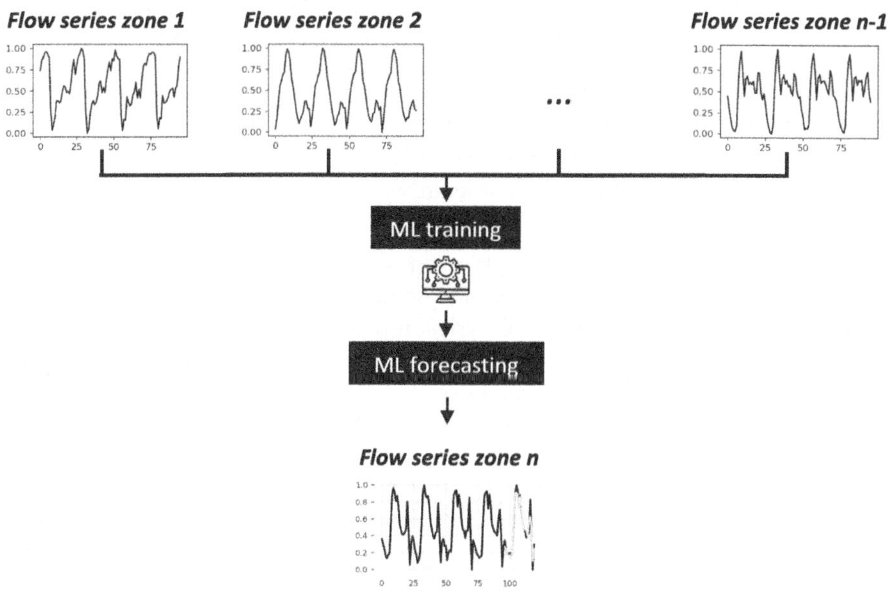

Fig. 3. Water demand forecasting for one zone through a ML model trained with all the flowrates series excluding the forecasted one [21].

3.2 N-Beats and Global Forecasting Model

This section delves into the utilization of the N-Beats model and explains the mechanism of training global models within the Darts library.

The N-Beats (Neural basis expansion analysis for interpretable time series forecasting) model, introduced by Oreshkin et al. in [46], was designed specifically for time series forecasting. The model architecture, depicted in Fig. 4, will be detailed herein; for a deeper understanding, additional insights can be found in [46].

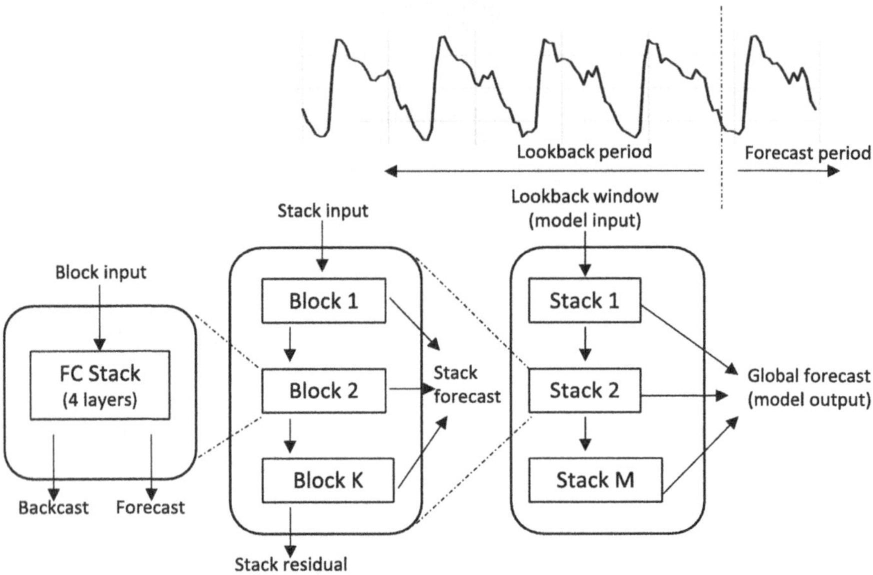

Fig. 4. N-Beats architecture [21].

In essence, considering a forecast horizon (H) and an observed series history (T), where T = n × H, the model employs the lookback period (observed series history) to understand the behavior of the time series. It predicts the behavior of the same time series in the forecast period (as shown in the upper part of Fig. 4). The model comprises multiple stacks (depicted in the right part of Fig. 4), with each stack containing several blocks (illustrated in the middle part of Fig. 4). Each block encompasses a fully connected stack with four layers responsible for both forecasting and backcasting (demonstrated in the left part of Fig. 4). The distinction between forecasting and backcasting lies in the direction of predictions: forecasting predicts future values by examining historical data, while backcasting extrapolates past values from future data, forecasting backward in time. The ReLU activation function introduces non-linearities within the model, aiding in learning complex patterns within the data. ReLU, a popular choice among activation functions due to its simplicity and speed [50], offers these essential non-linear properties crucial for resolving real-world problems.

This model, integrated into the Darts Python library, provides multiple forecasting models. However, only a subset, including N-Beats, can be trained with multiple time series. These models, termed global forecasting models, categorize time series into target time series (series for forecasting) and covariate time series (series considered to aid in forecasting target series). For this paper, the focus is solely on target series, as the historical water demand predominantly influences future demand [29, 51].

In scenarios where multiple target time series need training (as in this case), Darts constructs a dataset aggregating multiple input/output pairs from the provided time series. The input length corresponds to the 'input_chunk_length' hyperparameter, while the output length depends on the 'output_chunk_length' hyperparameter. Figure 5 illustrates the training phase using two example series of different lengths and timestamps for input. The number of training samples is determined by subtracting the sum of 'input_chunk_length' and 'output_chunk_length' from the time series length and adding 1 to the result. Consequently, the number of samples for training the first series equals 9, while the second series utilizes 7 samples. A training epoch in models handling multiple series involves a complete pass over all samples of each series.

Importantly, series need not have identical lengths, timestamps, or frequencies (although this does not apply to our specific case).

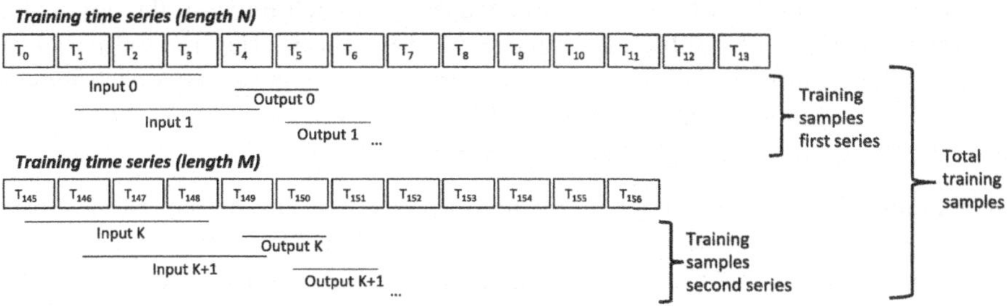

Fig. 5. Training of global models in Darts library with time series of different timestamps [21].

3.3 Performance Evaluation

Multi-step-ahead and one-step-ahead forecasting refer to different ways of predicting future values based on historical data. Both were implemented and compared using the N-Beats model.

Multi-step Ahead. In multi-step ahead forecasting multiple future values in a time series beyond the next single point are predicted. In other words, the model predicts several future points ahead in the time series horizon without using the predicted values as inputs for subsequent predictions.

One-Step Ahead. The one-step ahead forecasting involves predicting the value of the next point in a time series based on the available historical data up to the current point. This forecasting method is often used for real-time predictions or when the goal is to continuously update forecasts as new data becomes available.

For one-step ahead forecasting, a particular validation technique was proposed: the Walk Forward Validation (WFV). Time series data exhibit a distinctive characteristic known as autocorrelation, wherein observations close in time display correlation. Consequently, traditional cross-validation methods like K-fold or Shuffle split become inappropriate because they assume independence and identical distribution among samples. Essentially, these methods overlook the temporal relationships among observations during testing. To address this, a viable solution is employing a rolling-based cross-validation technique called Walk Forward Validation. This validation methodology involves training the model using the initial x observations and testing it against the subsequent x + 1 observation [52–54].

In response to this requirement, Darts offers two functions: 'historical_forecasts()' and 'backtest()'. The former function systematically constructs a training set in two ways: it either elongates the training set from the series' outset or maintains a fixed length specified as 'train_length'. Subsequently, the model is trained on this set. Following this, it generates a forecast of a length equal to 'forecast_horizon' and advances the end of the training set by 'stride' time steps. The 'start' parameter designates the beginning point for the forecast, set at 0.8 denoting that test starts at the first timestamp of the last 20% of the dataset. Therefore, dataset was divided into 80% training set and 20% test set. 'Forecast_horizon' signifies the number of time steps the model predicts into the future at each stage, configured as 1 for one-step-ahead forecasts. The 'stride' parameter, set at 1, dictates the number of time steps to advance after each forecast, thereby making successive predictions for subsequent time steps in the series. The 'retrain' parameter, marked as True, signifies the model's update with new data points after each forecast, triggering its retraining. On the other hand, the latter function, 'backtest()', directly furnishes the average error metric post-forecasting.

Performance Metrics. For both multi-step and one-step ahead forecasting, diverse error and performance metrics were used. The principal metric used to evaluate models is the Symmetric Mean Absolute Percentage Error (SMAPE), but other three metrics were selected to support the decision: Mean Absolute Error (MAE), Mean Squared Error (MSE), and Root Mean Square Error (RMSE) [45, 55–57].

Lower values of these metrics indicate better model performance overall.

4 Results and Discussion

This study introduces pioneering machine learning methodologies aimed at creating short-term forecasting models for water demand and energy. These methodologies leverage global forecasting models and transfer learning strategies to tackle this significant challenge.

4.1 Global and Local Models

Multi-step Ahead Forecasting. Initially, four distinct global models underwent testing to determine the most precise water demand forecasting model. The models assessed from the available options within the Darts Python library included N-Beats, RNN, BlockRNN, and Transformer.

Table 3 reports the average performances of each model.

Table 3. Performance metrics of N-Beats, RNN, BlockRNN, and Transformer global models [21].

Metrics	N-Beats	RNN	BlockRNN	Transf
MAE	0.031	0.188	0.23	0.16
SMAPE	21.322	52.15	56.174	45.762
MSE	0.003	0.06	0.075	0.043
RMSSE	0.043	0.237	0.265	0.194

Among the utilized models, N-Beats demonstrated superior performance across all evaluated metrics, affirming its superiority in forecasting accuracy. The remaining algorithms yielded notably similar metric outcomes, and the ranking based on performance, from highest to lowest, was as follows: N-Beats, Transformer, RNN, BlockRNN. For instance, considering the SMAPE metric, it is evident that the N-Beats model reduced this metric by over fifty percent compared to all other models. The exceptional performance of this model might be attributed to its capability to perform both backcasting and forecasting, a distinguishing feature highlighted before while elucidating its architecture.

Going deeply with the performance of N-Beats model, single metrics for each network zone are provided in Table 4.

Table 4. Performance metrics of N-Beats global model for each network zone.

	MAE	SMAPE	MSE	RMSE
Anfossi	0.041780	20.182580	0.003266	0.057148
Armi	0.011874	11.542808	0.000310	0.017616
Assiano	0.016780	13.637822	0.000530	0.023016
Baggio	0.018074	33.017148	0.000648	0.025457
Cantore	0.051254	19.405282	0.004831	0.069508
Chiusabella	0.015978	10.528552	0.000477	0.021851
Cimabue	0.029484	10.949384	0.001718	0.041452
Comasina	0.032597	11.243623	0.001811	0.042555
Crescenzago	0.039532	8.371949	0.002633	0.051316
Feltre	0.031648	13.038505	0.001659	0.040726
Gorla	0.014673	13.170710	0.000356	0.018876
Italia	0.064278	148.803420	0.014604	0.120848
Lambro	0.022586	17.617457	0.000944	0.030719
Novara	0.012297	12.238059	0.000208	0.014434

(continued)

Table 4. (*continued*)

	MAE	SMAPE	MSE	RMSE
Padova	0.026365	13.866038	0.001171	0.034220
S_Siro	0.028786	10.256632	0.001467	0.038297
Salemi	0.025349	13.049365	0.001214	0.034838
Suzzani_Nord	0.063090	25.221077	0.006844	0.082730
Suzzani_Sud	0.059324	14.596180	0.005155	0.071800
Vialba	0.011964	5.694689	0.000276	0.016601

Focusing on the SMAPE metric, among the 20 forecasts, 2 exhibited excellent results with a value below 10, 15 showed very good results with values below or nearly equal to 20, 2 forecasts fell between 25 and 35, while only 1 surpassed these values. Consequently, for the majority of the zones, the forecast proves to be highly accurate. Indeed, excluding the forecast with the highest error scores, the new average scores are MAE = 0.029, SMAPE = 14.612, MSE = 0.002, RMSE = 0.039, that are considerably lower than the previous ones.

The performance metrics of the N-Beats global forecasting model (presented in Table 4, first column) were compared to metrics derived from the construction of individual models trained separately with one series at a time. The following average values were obtained in the latter case: MAE = 0.030, SMAPE = 21.839, MSE = 0.003, RMSE = 0.043. Notably, nearly identical results were observed. This similarity might be attributed to the use of a limited amount of data; however, constructing a global forecasting model trained on a more extensive dataset over a longer duration could potentially enhance learning from diverse patterns simultaneously [58]. As indicated, related time series could significantly improve overall predictions compared to those obtained through a collection of local models.

Interestingly, the average time taken to train a single model with multiple time series (36 s) was notably less than the average aggregate time required to train individual models for each series (476 s). This difference can be justified by the fact that the complexity of local models increases proportionally with the number of time series in the dataset, surpassing the constant complexity of the global model [58].

As previously mentioned in the introduction, during the Water 4.0 era, water distribution systems have witnessed increased integration of sensors, Internet of Things (IoT), and large-scale data. Consequently, there's a wealth of data available for collection, analysis, and utilization in decision-making and planning processes. However, effectively managing this abundance of information is a crucial aspect. The outcomes obtained highlight that employing global models aligns with these requirements, as forecasting can be accomplished by training a single model on multiple series, requiring less time than the creation of numerous local models.

Table 5 reports performance metrics of local models for each zone.

Table 5. Performance metrics of N-Beats local model for each network zone.

	MAE	SMAPE	MSE	RMSE
Anfossi	0.041339	19.845489	0.003201	0.056577
Armi	0.010546	11.247974	0.000238	0.015440
Assiano	0.016679	12.561351	0.000538	0.023199
Baggio	0.019006	33.446855	0.000693	0.026332
Cantore	0.040660	16.789167	0.003240	0.056917
Chiusabella	0.015575	10.889078	0.000466	0.021585
Cimabue	0.025678	11.732680	0.001219	0.034908
Comasina	0.029965	12.958351	0.001685	0.041047
Crescenzago	0.041679	15.536962	0.003027	0.055020
Feltre	0.031272	12.862094	0.001726	0.041546
Gorla	0.012890	10.790274	0.000297	0.017228
Italia	0.059625	149.972923	0.016140	0.127043
Lambro	0.025212	23.091272	0.001178	0.034316
Novara	0.012337	12.138141	0.000278	0.016668
Padova	0.023145	11.452034	0.001038	0.032220
S_Siro	0.031189	12.484137	0.001811	0.042559
Salemi	0.027804	12.362763	0.001418	0.037651
Suzzani_Nord	0.066503	25.982124	0.008327	0.091254
Suzzani_Sud	0.054029	14.870077	0.004900	0.069998
Vialba	0.010401	5.769968	0.000211	0.014533

Table 5 presents almost similar SMAPE results of Table 4, with 1 forecast with a value less than 10, 15 showed very good results with values below or nearly equal to 20, 1 forecast between 20 and 25, 2 forecasts fell between 25 and 35, while only 1 surpassed these values. As before, excluding the time series with the highest scores, the new metrics are lower than the previous ones (MAE = 0.028; SMAPE = 15.095; MSE = 0.002; RMSE = 0.038).

One-Step Ahead Forecasting. The same model was tested in a one-step-ahead forecasting scenario to evaluate its ability to predict the immediate subsequent hour rather than forecasting for the entire following 24-h period. These average values were obtained: MAE = 0.037, SMAPE = 23.4, MSE = 0.004, RMSE = 0.037. The first three metrics are greater but almost equal to multi-step-ahead model results, but a lower RMSE is detected in this case.

Table 6 reports performance metrics for each zone.

Table 6. Performance metrics of N-Beats global model for each network zone

	MAE	SMAPE	MSE	RMSE
Anfossi	0.044067	19.313626	0.003256	0.044067
Armi	0.025640	25.646225	0.004067	0.025640
Assiano	0.024935	14.495221	0.001633	0.024935
Baggio	0.032622	37.871904	0.004928	0.032622
Cantore	0.050321	19.424105	0.004141	0.050321
Chiusabella	0.020775	9.978806	0.001238	0.020775
Cimabue	0.025519	10.313431	0.001152	0.025519
Comasina	0.029977	11.368504	0.001504	0.029977
Crescenzago	0.062470	16.918591	0.009491	0.062470
Feltre	0.035078	11.489443	0.001941	0.035078
Gorla	0.019046	9.777366	0.001278	0.019046
Italia	0.078079	155.061928	0.019381	0.078079
Lambro	0.027953	24.394969	0.001246	0.027953
Novara	0.027154	21.780178	0.001420	0.027154
Padova	0.043416	15.728827	0.006498	0.043416
S_Siro	0.033682	10.772451	0.001810	0.033682
Salemi	0.027581	7.379754	0.001502	0.027581
Suzzani_Nord	0.057542	25.999967	0.005756	0.057542
Suzzani_Sud	0.051310	15.484379	0.004056	0.051310
Vialba	0.016025	4.809697	0.000501	0.016025

As before, the new average metrics obtained subtracting the zone with the highest forecast are MAE = 0.034, SMAPE = 16.471, MSE = 0.003, RMSE = 0.034.

Hence, the multi-step-ahead forecasting model demonstrates higher accuracy compared to the one-step-ahead model. This superiority might be attributed to the multi-step forecasting model's enhanced capability to capture longer-term patterns. By predicting multiple future steps simultaneously, these models efficiently incorporate broader trends and seasonal variations present in the data, leading to more precise predictions over extended forecasting horizons.

Therefore, after confirming that the model yielded superior results with multi-step forecasting compared to one-step forecasting, the subsequent experiment results only consider the outcomes obtained from multi-step ahead forecasting.

4.2 Forecasting on Unseen Series

Water Demand Forecasting. The potential of global forecasting models to predict previously unobserved series was explored. For each flowrate time series, a model was created by training it with all other series and subsequently testing its performance on the excluded one.

Among 20 forecasts conducted, 18 exhibited highly satisfactory results, showcasing the models' strong ability to generalize effectively (MAE = 0.079; SMAPE = 25.141; MSE = 0.012; RMSE = 0.1). However, the remaining two forecasts were deemed unacceptable (MAE = 0.203; SMAPE = 97.746; MSE = 0.057; RMSE = 0.238).

The poorer performance observed in these specific cases within this limited group of time series could be attributed to their distinct patterns diverging significantly from those present in the group of time series used for model training. Consequently, the presence of two series with entirely different patterns than the others suggests that a global model comprising as diverse and heterogeneous series as possible could potentially yield better performance, particularly in scenarios involving transfer learning.

The performance metrics for each zone are summarized in Table 7.

Figure 6 shows the forecasting results of global model for three zones (Feltre, Cimabue, and Comasina) and the results of forecasting using the model trained with all flowrate time series excluding the forecasted one.

Table 7. Performance metrics of N-Beats global model for unseen network zones

	MAE	SMAPE	MSE	RMSE
Anfossi	0.145119	38.693579	0.038543	0.196324
Armi	0.137021	41.585159	0.030410	0.174383
Assiano	0.093374	33.564545	0.011058	0.105156
Baggio	0.098393	52.962700	0.017370	0.131796
Cantore	0.093093	30.053328	0.013806	0.117497
Chiusabella	0.209906	45.963534	0.062892	0.250784
Cimabue	0.041742	12.681174	0.003760	0.061322
Comasina	0.037777	9.855658	0.002511	0.050106
Crescenzago	0.074688	18.076108	0.007861	0.088665
Feltre	0.067240	13.230107	0.005733	0.075716
Gorla	0.039839	18.757027	0.003051	0.055236
Italia	0.196090	149.529287	0.051124	0.226107
Lambro	0.034457	24.882643	0.001894	0.043524
Novara	0.066299	24.084913	0.007073	0.084101
Padova	0.062058	22.312552	0.005800	0.076159
S_Siro	0.104738	19.837118	0.017921	0.133869

(continued)

Table 7. (*continued*)

	MAE	SMAPE	MSE	RMSE
Salemi	0.070092	16.512114	0.006662	0.081620
Suzzani_Nord	0.067283	27.817938	0.007680	0.087636
Suzzani_Sud	0.072047	21.120266	0.007584	0.087087
Vialba	0.116202	26.514683	0.021876	0.147906

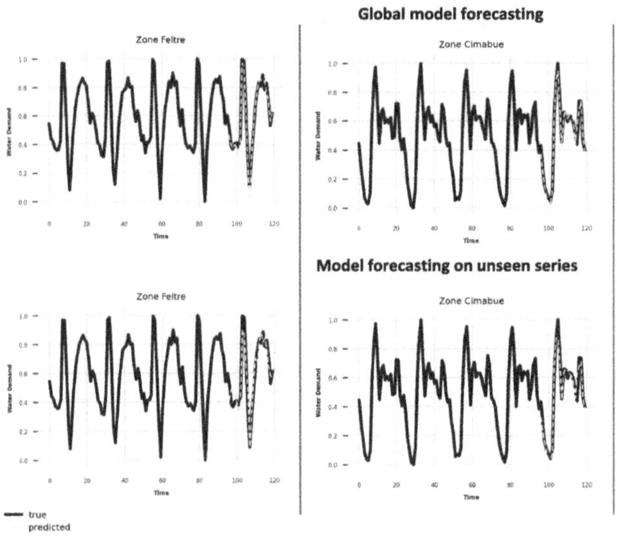

Fig. 6. Water demand forecasting for three zones with the global forecasting model, and forecasting results of the same zones with the model trained using all time series excluding the forecasted one [21]

Energy Consumption Forecasting. The energy consumption patterns of pumps are linked to the water demand within their respective zone. Consequently, the model previously trained on water demand time series was utilized to forecast pumps' energy consumption, aiming for this objective.

The resulting performance metrics validate the effectiveness of this approach (MAE = 0.032; SMAPE = 21.305, MSE = 0.003; RMSE = 0.043).

5 Conclusions

In this research, the authors devised a short-term forecasting method for water demand and pumps' energy consumption using simulated data from the Milan water distribution network. Specifically, they utilized hourly data to predict 24-h outcomes.

The approaches introduced in this study diverge from previous research in several significant aspects. Firstly, they examine both water and energy forecasts concurrently within the same water distribution network. Secondly, this study marks the inaugural use of global models within the water sector, facilitating the creation of a swift, comprehensive model capable of generalizing on unobserved time series (known as transfer learning).

Notably, the application of N-Beats in forecasting the water demand and pumps' energy consumption for an entire water distribution system represents a novel endeavor. The findings of this study indicate that N-Beats exhibited superior performance, suggesting its potential suitability for this domain.

Future investigations aim to validate this methodology using real data spanning an extended duration. Additionally, the objective is to construct more intricate models proficient in recognizing patterns related to weekdays, weekends, yearly trends, and potential long-term projections.

Acknowledgments. The research results presented in this paper have been achieved inside the Water 4.0 project, named "Technologies for the convergence between industry 4.0 and the integrated water cycle". This research project is currently running and is funded by the Ministry of Enterprises and Made in Italy (https://www.mimit.gov.it/en/).

Disclosure of Interests. The authors have no competing interests to declare that are relevant to the content of this article.

References

1. Hans, A., Bharat, D.A.: Water as a resource: different perspectives in literature. Int. J. Eng. Res. Technol. **3**, 27–34 (2014). https://doi.org/10.17577/IJERTV3IS100054
2. Patil, R., Alandikar, P., Chaudhari, V., Patil, P., Deshpande, S.: Water demand prediction using machine learning. IJRASET **10**, 122–128 (2022). https://doi.org/10.22214/ijraset.2022.47797
3. Leitão, J., Simões, N., Sá Marques, J.A., Gil, P., Ribeiro, B., Cardoso, A.: Detecting urban water consumption patterns: a time-series clustering approach. Water Supply **19**, 2323–2329 (2019). https://doi.org/10.2166/ws.2019.113
4. de Souza Groppo, G., Costa, M.A., Libânio, M.: Predicting water demand: a review of the methods employed and future possibilities. Water Supply **19**, 2179–2198 (2019). https://doi.org/10.2166/ws.2019.122
5. Esen, Ö., Yıldırım, D.Ç., Yıldırım, S.: Threshold effects of economic growth on water stress in the Eurozone. Environ. Sci. Pollut. Res. **27**, 31427–31438 (2020). https://doi.org/10.1007/s11356-020-09383-y
6. Hussain, Z., et al.: A comparative appraisal of classical and holistic water scarcity indicators. Water Resour. Manag. **36**, 931–950 (2022). https://doi.org/10.1007/s11269-022-03061-z
7. Stańczyk, J., Kajewska-Szkudlarek, J., Lipiński, P., Rychlikowski, P.: Improving short-term water demand forecasting using evolutionary algorithms. Sci. Rep. **12**, 13522 (2022). https://doi.org/10.1038/s41598-022-17177-0
8. Adedeji, K.B., Ponnle, A.A., Abu-Mahfouz, A.M., Kurien, A.M.: Towards digitalization of water supply systems for sustainable smart city development—water 4.0. Appl. Sci. **12**, 9174 (2022). https://doi.org/10.3390/app12189174
9. Sharif, M.N., Haider, H., Farahat, A., Hewage, K., Sadiq, R.: Water–energy nexus for water distribution systems: a literature review. Environ. Rev. **27**, 519–544 (2019). https://doi.org/10.1139/er-2018-0106

10. Wei, N., Li, C., Peng, X., Zeng, F., Lu, X.: Conventional models and artificial intelligence-based models for energy consumption forecasting: a review. J. Petrol. Sci. Eng. **181**, 106187 (2019). https://doi.org/10.1016/j.petrol.2019.106187
11. Mesalie, R.A., Aklog, D., Kifelew, M.S.: Failure assessment for drinking water distribution system in the case of Bahir Dar institute of technology, Ethiopia. Appl Water Sci **11**, 138 (2021). https://doi.org/10.1007/s13201-021-01465-7
12. Bolognesi, A., Bragalli, C., Lenzi, C., Artina, S.: Energy efficiency optimization in water distribution systems. Procedia Eng. **70**, 181–190 (2014). https://doi.org/10.1016/j.proeng.2014.02.021
13. Kofinas, D., Papageorgiou, E., Laspidou, C., Mellios, N., Kokkinos, K.: Daily multivariate forecasting of water demand in a touristic island with the use of artificial neural network and adaptive neuro-fuzzy inference system. In: 2016 International Workshop on Cyber-Physical Systems for Smart Water Networks (CySWater), pp. 37–42 (2016). https://doi.org/10.1109/CySWater.2016.7469061
14. Fares, A., Tijani, I.A., Rui, Z., Zayed, T.: Leak detection in real water distribution networks based on acoustic emission and machine learning. Environ. Technol. **44**, 3850–3866 (2023). https://doi.org/10.1080/09593330.2022.2074320
15. Berlotti, M., Di Grande, S., Cavalieri, S., Gueli, R.: Detection and prediction of leakages in water distribution networks. In: Proceedings of the 12th International Conference on Data Science, Technology and Applications, pp. 436–443. SCITEPRESS - Science and Technology Publications, Rome (2023). https://doi.org/10.5220/0012122000003541
16. Niknam, A., Zare, H.K., Hosseininasab, H., Mostafaeipour, A., Herrera, M.: A critical review of short-term water demand forecasting tools—what method should i use? Sustainability **14**, 5412 (2022). https://doi.org/10.3390/su14095412
17. Gan, X., Pei, J., Pavesi, G., Yuan, S., Wang, W.: Application of intelligent methods in energy efficiency enhancement of pump system: a review. Energy Rep. **8**, 11592–11606 (2022). https://doi.org/10.1016/j.egyr.2022.09.016
18. Kühnert, C., Gonuguntla, N.M., Krieg, H., Nowak, D., Thomas, J.A.: Application of LSTM networks for water demand prediction in optimal pump control. Water **13**, 644 (2021). https://doi.org/10.3390/w13050644
19. Ahmad, S., Jia, H., Chen, Z., Li, Q., Xu, C.: Water-energy nexus and energy efficiency: a systematic analysis of urban water systems. Renew. Sustain. Energy Rev. **134**, 110381 (2020). https://doi.org/10.1016/j.rser.2020.110381
20. Sarmas, E., Spiliotis, E., Marinakis, V., Tzanes, G., Kaldellis, J.K., Doukas, H.: ML-based energy management of water pumping systems for the application of peak shaving in small-scale islands. Sustain. Cities Soc. **82**, 103873 (2022). https://doi.org/10.1016/j.scs.2022.103873
21. Di Grande, S., Berlotti, M., Cavalieri, S., Gueli, R.: A proactive approach for the sustainable management of water distribution systems. In: Proceedings of the 12th International Conference on Data Science, Technology and Applications, pp. 115–125. SCITEPRESS - Science and Technology Publications, Rome (2023). https://doi.org/10.5220/0012121200003541
22. Yi, S., Kondolf, G.M., Sandoval-Solis, S., Dale, L.: Application of machine learning-based energy use forecasting for inter-basin water transfer project. Water Resour. Manag. **36**, 5675–5694 (2022). https://doi.org/10.1007/s11269-022-03326-7
23. Alhendi, A.A., et al.: Artificial intelligence for water–energy nexus demand forecasting: a review. Int. J. Low-Carbon Technol. **17**, 730–744 (2022). https://doi.org/10.1093/ijlct/ctac043
24. Ebrahim Banihabib, M., Mousavi-Mirkalaei, P.: Extended linear and non-linear auto-regressive models for forecasting the urban water consumption of a fast-growing city in an arid region. Sustain. Cities Soc. **48**, 101585 (2019). https://doi.org/10.1016/j.scs.2019.101585

25. Karamaziotis, P.I., Raptis, A., Nikolopoulos, K., Litsiou, K., Assimakopoulos, V.: An empirical investigation of water consumption forecasting methods. Int. J. Forecast. **36**, 588–606 (2020). https://doi.org/10.1016/j.ijforecast.2019.07.009
26. Ristow, D.C.M., Henning, E., Kalbusch, A., Petersen, C.E.: Models for forecasting water demand using time series analysis: a case study in Southern Brazil. J. Water Sanitation Hyg. Dev. **11**, 231–240 (2021). https://doi.org/10.2166/washdev.2021.208
27. Guo, G., Liu, S., Wu, Y., Li, J., Zhou, R., Zhu, X.: Short-term water demand forecast based on deep learning method. J. Water Resour. Plann. Manag. **144**, 04018076 (2018). https://doi.org/10.1061/(ASCE)WR.1943-5452.0000992
28. Salloom, T., Kaynak, O., He, W.: A novel deep neural network architecture for real-time water demand forecasting. J. Hydrol. **599**, 126353 (2021). https://doi.org/10.1016/j.jhydrol.2021.126353
29. Hu, S., Gao, J., Zhong, D., Deng, L., Ou, C., Xin, P.: An innovative hourly water demand forecasting preprocessing framework with local outlier correction and adaptive decomposition techniques. Water **13**, 582 (2021). https://doi.org/10.3390/w13050582
30. Shabani, S., Yousefi, P., Naser, G.: Support vector machines in urban water demand forecasting using phase space reconstruction. Procedia Eng. **186**, 537–543 (2017). https://doi.org/10.1016/j.proeng.2017.03.267
31. Herrera, M., Torgo, L., Izquierdo, J., Pérez-García, R.: Predictive models for forecasting hourly urban water demand. J. Hydrol. **387**, 141–150 (2010). https://doi.org/10.1016/j.jhydrol.2010.04.005
32. Brentan, B.M., Luvizotto, E., Jr., Herrera, M., Izquierdo, J., Pérez-García, R.: Hybrid regression model for near real-time urban water demand forecasting. J. Comput. Appl. Math. **309**, 532–541 (2017). https://doi.org/10.1016/j.cam.2016.02.009
33. Candelieri, A., et al.: Tuning hyperparameters of a SVM-based water demand forecasting system through parallel global optimization. Comput. Oper. Res. **106**, 202–209 (2019). https://doi.org/10.1016/j.cor.2018.01.013
34. Sardinha-Lourenço, A., Andrade-Campos, A., Antunes, A., Oliveira, M.S.: Increased performance in the short-term water demand forecasting through the use of a parallel adaptive weighting strategy. J. Hydrol. **558**, 392–404 (2018). https://doi.org/10.1016/j.jhydrol.2018.01.047
35. Herrera, M., García-Díaz, J.C., Izquierdo, J., Pérez-García, R.: Municipal water demand forecasting: tools for intervention time series. Stoch. Anal. Appl. **29**, 998–1007 (2011). https://doi.org/10.1080/07362994.2011.610161
36. Kofinas, D., Mellios, N., Papageorgiou, E., Laspidou, C.: Urban water demand forecasting for the island of skiathos. Procedia Eng. **89**, 1023–1030 (2014). https://doi.org/10.1016/j.proeng.2014.11.220
37. Li, Z., Zou, Z., Wang, L.: Analysis and forecasting of the energy consumption in wastewater treatment plant. Math. Probl. Eng. **2019**, e8690898 (2019). https://doi.org/10.1155/2019/8690898
38. Oliveira, P., Fernandes, B., Analide, C., Novais, P.: Forecasting energy consumption of wastewater treatment plants with a transfer learning approach for sustainable cities. Electronics **10**, 1149 (2021). https://doi.org/10.3390/electronics10101149
39. Bagherzadeh, F., Nouri, A.S., Mehrani, M.-J., Thennadil, S.: Prediction of energy consumption and evaluation of affecting factors in a full-scale WWTP using a machine learning approach. Process. Saf. Environ. Prot. **154**, 458–466 (2021). https://doi.org/10.1016/j.psep.2021.08.040
40. Shine, P., Murphy, M.D., Upton, J., Scully, T.: Machine-learning algorithms for predicting on-farm direct water and electricity consumption on pasture based dairy farms. Comput. Electron. Agric. **150**, 74–87 (2018). https://doi.org/10.1016/j.compag.2018.03.023

41. Wei, N., et al.: Short-term forecasting of natural gas consumption using factor selection algorithm and optimized support vector regression. J. Energy Resour. Technol. **141**, 032701 (2018). https://doi.org/10.1115/1.4041413
42. He, W.: Load forecasting via deep neural networks. Procedia Comput. Sci. **122**, 308–314 (2017). https://doi.org/10.1016/j.procs.2017.11.374
43. Wei, N., Li, C., Duan, J., Liu, J., Zeng, F.: Daily natural gas load forecasting based on a hybrid deep learning model. Energies **12**, 218 (2019). https://doi.org/10.3390/en12020218
44. Montero-Manso, P., Hyndman, R.J.: Principles and algorithms for forecasting groups of time series: locality and globality. Int. J. Forecast. **37**, 1632–1653 (2021). https://doi.org/10.1016/j.ijforecast.2021.03.004
45. Bandara, K., Hewamalage, H., Liu, Y.-H., Kang, Y., Bergmeir, C.: Improving the accuracy of global forecasting models using time series data augmentation. Pattern Recogn. **120**, 108148 (2021). https://doi.org/10.1016/j.patcog.2021.108148
46. Oreshkin, B.N., Carpov, D., Chapados, N., Bengio, Y.: N-BEATS: neural basis expansion analysis for interpretable time series forecasting (2020). https://doi.org/10.48550/arXiv.1905.10437. http://arxiv.org/abs/1905.10437
47. Mair, M., Rauch, W., Sitzenfrei, R.: Improving incomplete water distribution system data. Procedia Eng. **70**, 1055–1062 (2014). https://doi.org/10.1016/j.proeng.2014.02.117
48. Klise, K., et al.: Water Network Tool for Resilience (WNTR). User Manual, Version 0.2.3. Sandia National Lab. (SNL-NM), Albuquerque (2020). https://doi.org/10.2172/1660790
49. Herzen, J., et al.: Darts: user-friendly modern machine learning for time series. J. Mach. Learn. Res. **23**, 124:5442–124:5447 (2023)
50. Nair, V., Hinton, G.E.: Rectified linear units improve restricted boltzmann machines (2010)
51. Bakker, M., Vreeburg, J.H.G., van Schagen, K.M., Rietveld, L.C.: A fully adaptive forecasting model for short-term drinking water demand. Environ Model Softw. **48**, 141–151 (2013). https://doi.org/10.1016/j.envsoft.2013.06.012
52. Barzola-Monteses, J., Gómez-Romero, J., Espinoza-Andaluz, M., Fajardo, W.: Hydropower production prediction using artificial neural networks: an Ecuadorian application case. Neural Comput. Appl. **34**, 13253–13266 (2022). https://doi.org/10.1007/s00521-021-06746-5
53. Ngoc, T.T., Dai, L.V., Phuc, D.T.: Grid search of multilayer perceptron based on the walk-forward validation methodology. IJECE **11**, 1742 (2021). https://doi.org/10.11591/ijece.v11i2.pp1742-1751
54. Bergmeir, C., Benítez, J.M.: On the use of cross-validation for time series predictor evaluation. Inf. Sci. **191**, 192–213 (2012). https://doi.org/10.1016/j.ins.2011.12.028
55. Koutsandreas, D., Spiliotis, E., Petropoulos, F., Assimakopoulos, V.: On the selection of forecasting accuracy measures. J. Oper. Res. Soc. **73**, 937–954 (2022). https://doi.org/10.1080/01605682.2021.1892464
56. Botchkarev, A.: Performance metrics (error measures) in machine learning regression, forecasting and prognostics: properties and typology. IJIKM **14**, 045–076 (2019). https://doi.org/10.28945/4184
57. Hyndman, R.J., Koehler, A.B.: Another look at measures of forecast accuracy. Int. J. Forecast. **22**, 679–688 (2006). https://doi.org/10.1016/j.ijforecast.2006.03.001
58. Hewamalage, H., Bergmeir, C., Bandara, K.: Global models for time series forecasting: a simulation study. Pattern Recogn. **124**, 108441 (2022). https://doi.org/10.1016/j.patcog.2021.108441

Disaster Tweet Classification Using Fine-Tuned Deep Learning Models Versus Zero and Few-Shot Large Language Models

Soudabeh Taghian Dinani(✉), Doina Caragea, and Nikesh Gyawali

Kansas State University, Manhattan, KS, USA
{soudabehtaghian,dcaragea,gnikesh}@ksu.edu

Abstract. The task of effectively filtering and categorizing the vast volume of user-generated content on social media during disaster events is crucial for emergency management and disaster response organizations to allocate their resources efficiently. Deep learning approaches, including recurrent neural networks, Capsule Neural Networks (CapsNets) and transformer-based models, have been successfully used for text classification. However, they have not been extensively compared in the context of crisis tweet classification. Furthermore, such models have not been compared with foundational large language models (LLMs), such as the open-source Llama 2, which are known to produce good results in zero and few-shot settings. In this study, we conducted a comprehensive comparison between CapsNet models, BERT models, and Bi-LSTM, one of the widely used recurrent neural network models, in the task of classifying crisis tweets. We also compared these models with Llama 2 in 0-shot, 1-shot, and 5-shot learning settings. We considered two classification tasks: binary classification for informativeness and multi-class classification for humanitarian content. For this purpose, we used several benchmark datasets for crisis tweet classification, namely CrisisBench, CrisisNLP, and CrisisLex. Our experiments revealed that the results of the CapsNet models are comparable to those of the Bi-LSTM models across various metrics. However, BERT models consistently outperformed both Bi-LSTM and CapsNet models across different datasets for both classification tasks. Notably, Llama 2 exhibited promising performance, considering its exposure to no data instances in the 0-shot and only few instances in 1-shot and 5-shot learning settings, showcasing its potential for handling crisis tweet classification. However, it is important to highlight that the fine-tuned models, including CapsNet, BERT, and Bi-LSTM, achieved superior performance overall, emphasizing the influence of fine-tuning on model effectiveness. These findings suggest that BERT is still one of the most effective models for classifying crisis tweets. Additionally, Llama 2, with its strong performance in limited-shot scenarios that match the scenarios in the crisis domain, presents an intriguing path for further exploration in handling crisis-related content.

Keywords: Tweet classification · Capsule neural networks · BERT · Bi-LSTM · Large language models · Llama 2

O. Gusikhin et al. (Eds.): DATA 2023, CCIS 2105, pp. 73–94, 2024.
https://doi.org/10.1007/978-3-031-68919-2_4

1 Introduction

Social media holds a crucial role in modern life, transforming virtual communication practices [52]. Social media, recognized for its rapid news distribution and facilitation of two-way communication, has progressively become a dominant medium for human interaction and communication [52]. Consequently, it is not surprising that individuals increasingly turn to social media during disaster events (such as earthquakes, hurricanes, fires, etc.), utilizing platforms to connect with family, seek emotional support, gather information on essentials like food and shelter, and share news [2]. This shift is particularly crucial when standard communication channels, such as emergency lines (such as 911 lines), may be overwhelmed due to the large number of calls received during disasters [49]. For example, during the floods caused by Hurricane Harvey in Texas in August 2017, unable to reach emergency lines, a lot of trapped people started posting messages on Twitter pleading for help. A user tweeted "Please help I'm stranded with my kids I need help fast," on the Houston Police's Twitter account. Another user tweeted "Our apartments are surrounded with water like an island we need rescue", and a baby was saved after the following message, along with a picture of a sleeping baby, was posted "Please help us she is a newborn" [25]. These social media posts provide crucial information for response organizations as they offer immediate, eyewitness-shared information during ongoing disasters, contributing to potential life-saving efforts [42].

Numerous studies have extensively highlighted the utility of information derived from social media platforms for disaster response and management [18]. However, the information derived from social media is often noisy and contains irrelevant content, particularly due to the high volume of tweets generated during unfolding disasters [4]. Additionally, manually extracting and processing pertinent information from social media is challenging, time-consuming, and costly, given the vast volume and speed of streaming data [42]. As a result, there is a need for computational techniques to effectively filter out relevant posts from this noisy data, enabling emergency responders to obtain actionable and situational information to aid the affected population [4].

Some previous studies employed traditional machine learning (ML) methods to analyze crisis-related social media data. The objective was to aid emergency responders in identifying and extracting pertinent posts, thereby improving relief operations through the identification of crisis-relevant information [16,34]. More recent studies have used deep learning (DL) methods for this purpose [3,20,22,27,28,42,45]. Specifically, transformer-based models, such as BERT, have demonstrated effectiveness in the classification of disaster tweets. In a recent investigation [5], BERT-type models outperformed convolutional neural networks (CNN) and FastText models [19], establishing themselves as state-of-the-art models for disaster tweet classification.

Another recently developed deep learning model that exhibits considerable promise in text classification is the CapsNet model. Initially designed for image classification [43], this model has demonstrated superior performance compared to baseline models like CNN, SVM-based models, and LSTM-based models, particularly in tasks such as sentiment classification [50,54,56], question categorization [56], question answering [55], news categorization [56], and text classification [8]. The superiority of CapsNet models has been demonstrated using various capsule architectures, including NLP-Capsule and CapsuleDAR [54]. In several studies, CapsNet models have exceeded baselines on multiple datasets and have proven their capability in transferring information from single-label to multi-label text classification [56].

Given that CapsNet models have shown promising results for disaster image classification [12], due to their hierarchical structure preserving spatial relationships between features and given that they have also been effective for different text classification tasks, we recently explored [11] if CapsNets can also benefit crisis-related tweet classification tasks. Specifically, we compared CapsNets (with static and dynamic routing algorithms) with the state-of-the-art BERT model and also LSTM/Bi-LSTM models on the tasks of classifying crisis-related tweets in terms of both their informativeness with respect to disaster response (binary classification) and their humanitarian content (multi-class classification). For this purpose, we performed an comprehensive set of experiments using three datasets: CrisisLex, CrisisNLP, and CrisisBench, with the last one being a benchmark dataset consolidating eight prior datasets. In this paper, we present an extended analysis building upon our prior work [11], where we compare three fine-tuned models, selected based on the prior results with a foundational LLM in 0-shot, 1-shot, and 5-shot learning settings. Specifically, as fine-tuned models, we use BERT (as it was shown to have best results overall in our prior work), CapsNet with static routing (as both static and dynamic routing were shown to give comparable results) and Bi-LSTM (which had slightly better performance than the standard LSTM). As for the foundational LLM, we use the open-source Llama 2 in 0-shot, 1-shot, and 5-shot settings for both informativeness and humanitarian content tasks.

Our extended comparison shows that the fine-tuned BERT model is still the best model for the tasks considered. However, Llama 2 shows very promising results, considering that it is used in a low-resource setting similar to what is encountered in practical scenarios in the crisis domain, making it a strong candidate when no labeled data is available for an emerging crisis event and real-time data filtering is needed.

The rest of the paper is structured as follows. Section 2 describes the methods used, specifically, Bi-LSTM, CapsNets, BERT and Llama 2. Sections 3 and 4 describe the implementation details (i.e., performance metrics, dataset, and experimental setup) and the result analysis, respectively. Section 5 discusses the related work. Finally, Sect. 6 concludes the paper.

2 Methods

In this section, we provide background for the methods used in this study, including Bidirectional LSTM (Bi-LSTM) networks, Bidirectional Encoder Representations from Transformers (BERT) networks, Capsule Networks (CapsNets) and Llama 2.

2.1 Bidirectional LSTM (Bi-LSTM)

Standard LSTM models, processing data in only one direction and relying solely on past information, may face challenges such as misinterpreting context or incomplete understanding [29]. In contrast, Bi-LSTM models [14], a variant of LSTM, can overcome this limitation by processing information bidirectionally. This approach involves simultaneous analysis of both past and future information/words, merging them for predicting the current output [31]. Utilizing two hidden LSTM layers-one forward and one backward-in its architecture, a Bi-LSTM model integrates the outputs of these layers to capture both forward and backward data streams within a sequence [29].

2.2 Bidirectional Encoder Representations from Transformers (BERT)

The Transformer model, incorporating self-attention [48], has demonstrated outstanding performance across various NLP applications [1]. The attention mechanism enables transformers to concurrently process words in a sequence, resulting in a more computationally efficient implementation suitable for training models on extensive corpora [48]. The advent of the transformer has led to the development of various transformer-based models, including BERT, for text encoding and classification tasks. BERT [10] is constructed on a multilayered bidirectional transformer encoder, in contrast to earlier contextual word embedding models, primarily unidirectional (e.g., Elmo) [21]. This bidirectional characteristic empowers BERT models to achieve state-of-the-art performance across diverse NLP tasks [21].

Devlin et al. [10] present two BERT architectures: BERT-Base and BERT-Large. The former comprises 12 layers (i.e., transformer blocks) with 12 self-attention heads, a hidden layer size of 768, and a total of 110 million parameters. In contrast, the latter consists of 24 layers (i.e., transformer blocks) with 16 self-attention heads, a hidden layer size of 1024, and a total of 340 million parameters.

Different variants of BERT, including RoBERTa (Robustly Optimized BERT pre-training Approach) [30] and DistilBERT [44], have been employed in previous studies for disaster tweet classification alongside the standard BERT model. However, Alam et al. [5] demonstrated that these three models-BERT, RoBERTa, and DistilBERT- exhibit comparable performance in tasks such as informativeness and humanitarian category classification. Consequently, we opted to use BERT as one of our baseline models.

2.3 Capsule Networks

Capsule Networks (CapsNets) originated in the field of image classification [43] as an innovative solution to address the limitations of Convolutional Neural Networks (CNNs). CNNs, which use subsampling/pooling layers, are prone to losing information, leading to translational invariance where neuron activity remains unchanged with viewpoint translation. This translational invariance results in the "Picasso problem", where CNNs attempt to identify an image by detecting its components without considering their spatial relationships. In contrast, CapsNets leverage the equivariant property, meaning that neuron activity changes with viewpoint variations, thereby preserving the spatial relationships between components [40].

A Capsule Network (CapsNet) comprises multiple layers of capsules, where each group of neurons is capable of capturing distinct properties of an object, such as its position or size. This design allows a capsule to have inputs and outputs represented as a vector (with the vector's length indicating the likelihood of an object) or a matrix, accompanied by an activation probability. In contrast, a conventional neural network restricts a single neuron to scalar values for both inputs and outputs [15,43].

A dynamic routing-by-agreement algorithm is employed to establish part-whole relationships between lower-layer capsules (parts) and higher-layer capsules (wholes). In this algorithm, if multiple predictions from active lower-layer capsules, determined through transformation matrices, are in agreement, a higher-layer capsule becomes active [43]. While Zhao et al. [56] showcased the effectiveness of CapsNets for text classification, Kim et al. [24] later proposed a static routing method for such tasks, demonstrating its superiority over dynamic routing. Consequently, in our study, we utilize the CapsNet architecture with static routing, as suggested by Kim et al. [24]. It is worth noting that, as our conference paper [11] indicated similar results for both dynamic and static routing, we solely present the results from static routing in this context.

2.4 Llama 2

Large Language Models (LLMs) represent a promising frontier as proficient AI assistants have been shown to be capable of executing intricate reasoning tasks across diverse domains [37]. Their swift and extensive adoption among the general public stems from their ability to facilitate human interaction through user-friendly chat interfaces [37].

Meta GenAI recently introduced Llama 2, an evolved iteration of Llama 1 [46]. This open-source Llama 2 variant, just pretrained and also fine-tuned, has undergone extensive evaluation and has been shown to be competitive with cutting-edge closed-source LLMs. The released Llama 2 family comprises pretrained and fine-tuned models, offered in three variants with 7B, 13B, and 70B parameters. Llama 2, functioning as an auto-regressive language model, incorporates an optimized transformer architecture. The tuned versions undergo supervised fine-tuning (SFT) and reinforcement learning with human feedback (RLHF) to align with human preferences for helpfulness [47]. These models are characterized by their extensive pretraining and expansive parameter size, yielding exceptional generalization capacities. This enables them to excel even in tasks for which they were not explicitly trained, showcasing the unique strengths of Llama 2 in zero- and few-shot learning scenarios.

In our experiments, we specifically utilized the 7B variant of Llama 2, concentrating on its performance in zero- and few-shot learning scenarios, with a focus on 0-shot text classification (predicting class labels for text instances from unfamiliar categories without the need for labeled training examples) and few-shot text classification (involving the model's exposure to only a few instances of labeled training examples for each class). This deliberate selection is aimed at gaining insights into the adaptability and effectiveness of Llama 2 across different levels of exposure to training instances.

3 Implementation Details

In this section, the datasets used in the experiments, along with the experimental setup, and evaluation metrics are presented. The purpose of the conducted experiments is to answer the following research questions which motivated this study:

- How do the fine-tuned CapsNet, Bi-LSTM and BERT models compare to each other and to Llama 2 model with zero- and few-shots, when used for classifying crisis-related tweets in terms of informativeness (binary classification) and humanitarian category (multi-class classification)?

- How do the 0-shot and few-shot Llama 2 results compare to each other when classifying crisis-related tweets in terms of informativeness and humanitarian category?

3.1 Performance Metrics

In the experiments, we compare fine-tuned CapsNet, Bi-LSTM, BERT models with 0-shot and 1-shot and 5-shot Llama 2 models on both binary (tweet informativeness) and multi-class (tweet humanitarian category) classification tasks. For both types of tasks, the comparison is performed using several standard evaluation metrics, specifically, Precision, Recall and F1-scores for each class and Weighted Precision, Weighted Recall and Weighted F1-scores for the overall performance of each model, shown as Avg row in the result tables.

3.2 Datasets

The datasets used in this study comprise CrisisLex, CrisisNLP, and CrisisBench, and their respective statistics can be found in Table 2. All three datasets are annotated for both informativeness and humanitarian category tasks. While the informativeness task across all datasets involves two classes, namely Informative and Not-informative, the humanitarian category task differs in the number of classes, with 6, 10, and 11 classes in CrisisLex, CrisisNLP, and CrisisBench datasets, respectively. The names of their classes together with the name abbreviations used in the paper are shown in Table 1. A brief description of each dataset is given below.

- **CrisisLex** is formed by merging two datasets, namely CrisisLexT26 and CrisisLexT6 [39]. The CrisisLexT26 dataset encompasses tweets from 26 disasters occurring in 2012 and 2013, while the CrisisLexT6 dataset includes tweets from 6 disasters that took place between October 2012 and July 2013 [5].
- **CrisisNLP** consists of tweets from 19 disasters which happened between years 2013 and 2015 [17].
- **CrisisBench** is a benchmark dataset created by Alam et al. [5]. This dataset is a consolidation of eight earlier datasets, namely CrisisLex, CrisisNLP, SWDM2013, ISCRAM2013, Disaster Response Data (DRD), Disasters on Social Media (DSM), CrisisMMD, and AIDR.

3.3 Experimental Setup: Fine-Tuned Deep Learning Models

Given the availability of train/test/dev subsets for the three datasets used in the study, the experiments were repeated three times using distinct seeds, and the results reported in the tables represent the average of these three runs.

The hyperparameters utilized in the experiments were fine-tuned based on the corresponding development (dev) datasets. In particular, for the Bi-LSTM experiments, the following hyperparameters were employed: a hidden size of 256, a maximum sequence length of 60, a learning rate of 5e−4, a dropout ratio of 0.1, weight decay of 0 (i.e., L2 regularization was not applied), a batch size of 40, and a maximum number of epochs set to 10.

Table 1. List of tasks and classes in each task and the corresponding abbreviations (Abbr.)

Task/class	Abbr.	Task/class	Abbr.
Humanitarian	Hum.	Informativeness	Inf.
Affected individual	AI	Informative	I
Caution and advice	CA	Not informative	NI
Displaced and evacuations	DE		
Donation and volunteering	DV		
Infrastructure and utilities damage	IUD		
Injured or dead people	IDP		
Missing and found people	MFP		
Not humanitarian	NH		
Requests or needs	RN		
Response efforts	RE		
Sympathy and support	SS		

In the BERT experiments, the BERT base uncased model was employed, featuring 12 layers (i.e., Transformer blocks), 12 self-attention heads, and a hidden size of 768. The hyperparameters utilized for these experiments included a maximum sequence length of 60, a learning rate of 2e-5, a dropout ratio of 0, weight decay of 0 (i.e., L2 regularization was not applied), a batch size of 40, and a maximum number of epochs set to 10.

The hyperparameters employed in the CapsNet with static routing experiments consisted of a maximum sequence length of 60, a learning rate set to 5e-5, a dropout ratio of 0.6, weight decay of 0 (i.e., L2 regularization was not applied), a batch size of 40, and a maximum number of epochs set to 50.

The Adam optimizer was employed for all fine-tuned experiments. Additionally, the number of epochs that yielded the highest accuracy on the development set in each fine-tuned experiment was utilized to evaluate performance on the test set.

3.4 Experimental Setup: Llama 2 in 0-Shot and Few-Shot Settings

In our experiments, we specifically utilized Pretrained Llama-2-7B, concentrating on its performance in 0-shot and few-shot learning scenarios, with a focus on crisis-related tweet classification. In 1-shot, one instance from each class was randomly selected as a labeled example, and for 5 shots, four more examples were randomly selected and added to that instance. For few shots, three different sets of random examples were selected, and the results were averaged over these three runs, while for 0-shot, the experiments were run only once. Generic prompts used for the humanitarian tasks are provided below:

0-shot:

```
The following tweet is related to a disaster.
What humanitarian class does it belong to?
Reply with only one of these options without writing your reasoning:
{list_all_classes_in_this_task}

{input_tweet}
```

Table 2. Statistics for Crisis Datasets: The dictionary of classes and the corresponding abbreviations are presented in Table 1.

Task Class	CrisisLex				CrisisNLP				CrisisBench			
Inf.	Train	Dev	Test	Total	Train	Dev	Test	Total	Train	Dev	Test	Total
I	25955	3727	7447	37129	14865	2182	4087	21134	65826	9594	18626	94046
NI	18862	2769	5384	27015	11298	1645	3119	16062	43970	6414	12469	62853
Total	44817	6496	12831	64144	26163	3827	7206	37196	109796	16008	31095	156899
Hum.	Train	Dev	Test	Total	Train	Dev	Test	Total	Train	Dev	Test	Total
AI	2125	317	601	3043	–	–	–	–	2454	367	693	3514
CA	912	143	247	1302	706	99	198	1003	2101	309	583	2993
DE	–	–	–	–	322	51	88	461	359	53	99	511
DV	1031	158	293	1482	1647	265	470	2382	5184	763	1453	7400
IUD	752	79	216	1047	1128	163	300	1591	3541	511	1004	5056
IDP	–	–	–	–	1235	165	362	1762	1945	271	561	2777
MFP	–	–	–	–	278	44	80	402	373	55	103	531
NH	18844	2748	5423	27015	11227	1686	3150	16063	36109	5270	10256	51635
RN	–	–	–	–	103	19	29	151	4840	705	1372	6917
RE	–	–	–	–	780	113	221	1114	780	113	221	1114
SS	1800	283	532	2615	1624	237	455	2316	3549	540	1020	5109
Total	25464	3728	7312	36504	19050	2842	5353	27245	61235	8957	17365	87557

Few-shots:

```
You are an expert disaster-related tweet classifier with the ability
to distinguish humanitarian categories of tweets related to disasters.
Your task is to classify the given tweets into one of these
{number_of_classes} humanitarian categories: {list_all_classes_in_this_task}

Here are a few classification examples:
1) {tweet_example_1}
Category: {its_category}

2) {tweet_example_2}
Category: {its_category}

i) {tweet_example_i}
Category: {its_category}
...
Please classify each tweet into one of {number_of_classes} humanitarian categories.
Provide your answer as only one humanitarian category without writing your reasoning.

{input_tweet}
```

4 Experimental Results and Discussions

The experimental results of crisis tweet classification for informativeness category task of all three datasets are presented in Table 3 and for humanitarian category task of CrisisLex, CrisisNLP, and CrisisBench are presented in Tables 4, 5, and 6, respectively. The

Table 3. Experiment results for tweet informativeness task for three datasets of CrisisNLP, CrisisLex, and CrisisBench for each class and the averaged weighted metrics (Avg), with the best value of each metric displayed in **bold** within each row.

Dataset	Class	Metric	Models					
						Llama 2		
			Bi-LSTM	BERT	CapsNet	0-shot	1-shot	5-shot
CrisisLex	I	Pr	**94.746**	94.255	94.469	67.684	58.706	60.530
		Re	95.815	96.567	95.108	97.435	**99.324**	99.284
		F1	95.275	**95.393**	94.785	79.879	73.794	75.198
	NI	Pr	94.127	**95.101**	93.175	90.922	76.220	89.715
		Re	**92.642**	91.846	92.292	35.577	3.177	10.273
		F1	93.374	**93.436**	92.728	51.143	6.077	18.062
	Avg	Pr	94.486	**94.610**	93.926	77.428	66.047	72.772
		Re	94.484	**94.586**	93.926	71.497	59.024	61.950
		F1	94.477	**94.571**	93.922	67.829	45.411	51.234
CrisisNLP	I	Pr	85.099	**88.877**	84.075	58.514	58.596	58.928
		Re	86.747	85.825	85.735	97.895	**98.671**	98.621
		F1	85.879	**87.297**	84.888	73.247	73.511	73.771
	NI	Pr	82.256	**82.279**	80.823	76.567	83.472	84.524
		Re	79.994	**85.861**	78.690	9.015	8.477	9.897
		F1	81.042	**83.993**	79.725	16.131	14.854	17.593
	Avg	Pr	83.868	**86.022**	82.667	66.326	69.362	70.007
		Re	83.824	**85.840**	82.686	59.434	59.636	60.217
		F1	83.786	**85.867**	82.654	48.531	48.125	49.455
CrisisBench	I	Pr	88.036	**89.452**	87.916	65.167	60.407	61.447
		Re	90.543	90.708	89.622	97.213	**99.442**	99.026
		F1	89.270	**90.057**	88.755	78.028	75.158	75.836
	NI	Pr	85.244	**85.872**	84.041	84.279	75.521	83.314
		Re	81.601	**83.956**	81.572	22.334	2.540	7.110
		F1	83.377	**84.862**	82.776	35.311	4.913	13.072
	Avg	Pr	86.917	**88.016**	86.362	72.828	66.464	70.211
		Re	86.959	**88.002**	86.395	67.198	60.608	62.186
		F1	86.908	**87.975**	86.358	60.905	47.007	50.680

tables depict the performance of the models for each class (namely, Precision, Recall, and F1-score) in the datasets, along with the average performance of the models in the "Avg" row, which shows the Weighted Precision, Weighted Recall, and Weighted F1-score across all classes of each dataset. Moreover, to facilitate the comparison between various models, we have created two graphs that illustrate the F1-scores for different classes and the overall performance averages across all classes within a given task and

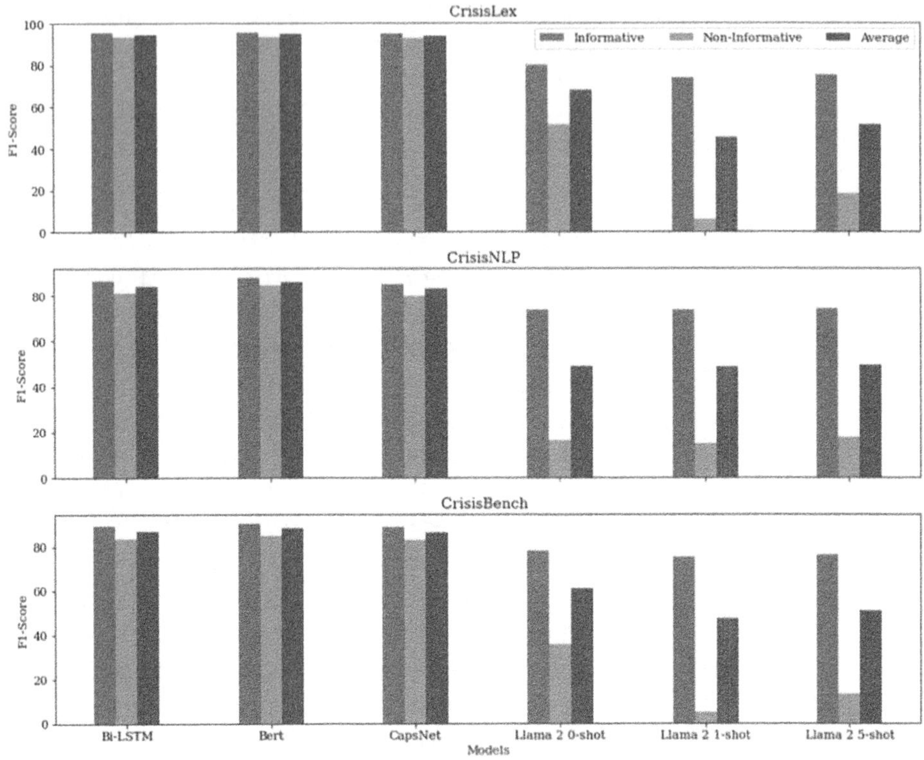

Fig. 1. The graph for F1-score comparison of Informativeness task for the three datasets, CrisisNLP, CrisisBench, respectively.

dataset. The graph depicting informativeness task across all three datasets is presented in Fig. 1, while the one for the humanitarian task is showcased in Fig. 2. We discuss the answers to our questions using these results in the following paragraphs.

Our first research question aims to evaluate the performance of fine-tuned deep learning models, including BERT, CapsNet, and Bi-LSTM against each other, and also in comparison to results obtained from the Llama 2 model utilizing zero- and few-shot settings. This assessment focused on their effectiveness in classifying crisis-related tweets concerning informativeness (binary classification) and humanitarian category (multi-class classification) across three datasets: CrisisLex, CrisisNLP, and CrisisBench. Our second question aims to assess and compare the results of the Llama 2 model in various settings, specifically, 0-shot, 1-shot and 5-shot, for both tasks.

For each task, we initially analyze the overall performance of the models, "Avg" rows in the tables, and then, examine the results for each class in the dataset. As can be seen in Tables 3, 4, 5, and 6, along with the corresponding F1-score graphs in Fig. 1 and Fig. 2, BERT models consistently outperform Bi-LSTM, CapsNet, and Llama 2 models (in all settings) for both informativeness and humanitarian tasks across all considered evaluation metrics.

Table 4. Experiment results for tweet humanitarian task for the CrisisLex dataset. Performance is reported in terms of Precision (Pr), Recall (Re) and F1-score (F1) for each class and also as a weighted average over all classes (Avg), with the best value of each metric displayed in **bold** within each row.

Class	Metric	Models					
					Llama 2		
		Bi-LSTM	BERT	CapsNet	0-shot	1-shot	5-shot
AI	Pr	83.053	**85.221**	82.065	19.951	44.547	53.552
	Re	81.697	**86.023**	83.250	53.910	27.843	16.528
	F1	82.358	**85.603**	82.632	29.124	33.375	20.406
CA	Pr	66.500	**72.484**	67.972	0.000	31.137	22.811
	Re	59.784	**71.660**	64.642	0.000	38.327	13.225
	F1	62.956	**72.042**	66.257	0.000	30.089	15.596
DV	Pr	76.660	**81.233**	78.492	7.582	53.348	81.118
	Re	76.678	**77.702**	75.313	88.055	57.565	19.113
	F1	76.624	**79.408**	76.832	13.961	52.525	27.675
IUD	Pr	56.238	**66.348**	61.407	0.000	54.738	24.040
	Re	**62.037**	58.642	46.450	0.000	9.105	5.401
	F1	58.378	**62.140**	52.864	0.000	15.423	8.660
NH	Pr	97.308	**97.543**	96.555	93.394	88.207	79.017
	Re	97.855	97.861	**97.984**	37.858	85.580	98.776
	F1	97.577	**97.702**	97.264	53.877	86.553	87.682
SS	Pr	81.578	**82.079**	79.516	38.636	41.746	74.540
	Re	77.882	**84.586**	77.506	3.208	65.880	18.734
	F1	79.581	**83.276**	78.464	5.923	50.725	26.780
Avg	Pr	91.911	**92.984**	91.500	73.911	76.893	73.156
	Re	91.881	**93.071**	91.808	36.270	74.389	77.336
	F1	91.850	**93.008**	91.609	43.287	74.167	70.534

For the tweet informativeness task, the F1-scores of BERT models are 94.571, 85.867, and 87.975 for CrisisLex, CrisisNLP, and CrisisBench datasets, respectively. These scores are higher than those of the other five models for the corresponding datasets. For this classification task and for the two datasets of CrisisLex, CrisisBench, the order of other models, from best to worst, is Bi-LSTM, CapsNet, Llama 2 0-shot, 5-shot, and 1-shot (with F1-scores of 94.477, 93.922, 67.829, 51.234, and 45.411 respectively, for CrisisLex dataset and 86.908, 86.358, 60.905, 50.680, and 47.007 respectively for CrisisBench dataset). For the CrisisNLP dataset, the order is Bi-LSTM, CapsNet, Llama 2 5-shot, 0-shot, and 1-shot (with F1-scores of 83.786, 82.654, 49.455, 48.531, and 48.125, respectively).

Regarding the performance on the classes of this task, the BERT model still exhibits superior F1-scores across all three datasets, with values of 95.393, 87.297, and 90.057

Table 5. Experiment results for tweet humanitarian task for the CrisisNLP dataset. Performance is reported in terms of Precision (Pr), Recall (Re) and F1-score (F1) for each class and also as a weighted average over all classes (Avg), with the best value of each metric displayed in **bold** within each row.

Class	Metric	Models					
					Llama 2		
		Bi-LSTM	BERT	CapsNet	0-shot	1-shot	5-shot
CA	Pr	69.971	**77.631**	66.060	4.167	26.646	12.649
	Re	**62.626**	57.407	56.061	0.505	13.299	0.842
	F1	65.862	**65.869**	60.364	0.901	17.645	1.520
DE	Pr	55.889	**64.369**	61.226	3.322	13.718	11.034
	Re	41.288	**59.470**	29.925	34.091	31.061	10.985
	F1	46.983	**61.188**	39.650	6.054	16.010	10.292
DV	Pr	61.887	**78.905**	64.779	12.459	42.710	52.955
	Re	**77.873**	73.688	70.142	88.511	7.518	3.759
	F1	68.824	**76.007**	67.285	21.843	12.706	6.282
IUD	Pr	75.072	**76.910**	76.248	0.000	68.939	33.333
	Re	62.667	**67.000**	61.667	0.000	15.333	0.111
	F1	67.970	**71.473**	67.924	0.000	24.338	0.221
IDP	Pr	83.200	**86.100**	82.294	40.000	60.725	33.333
	Re	86.464	**88.674**	80.019	21.170	11.520	0.276
	F1	84.785	**87.328**	81.093	27.687	19.122	0.548
MFP	Pr	44.762	**55.407**	54.939	17.241	40.863	0.000
	Re	39.583	**42.083**	33.750	6.250	10.485	0.000
	F1	41.949	**47.798**	41.606	9.174	16.581	0.000
NH	Pr	84.561	**85.820**	81.686	85.577	70.495	62.442
	Re	91.566	**94.656**	92.857	17.055	74.510	92.894
	F1	87.923	**90.005**	86.912	28.442	72.287	74.599
RN	Pr	2.564	**45.707**	0.000	1.449	2.818	16.667
	Re	1.149	**13.793**	0.000	3.448	4.598	2.299
	F1	1.587	**20.880**	0.000	2.041	3.391	4.040
RE	Pr	40.133	**49.081**	42.209	7.812	10.304	14.156
	Re	17.496	**37.255**	8.145	4.525	29.563	4.072
	F1	23.719	**41.376**	12.472	5.731	14.085	6.265
SS	Pr	74.843	**81.522**	63.531	16.000	30.464	10.812
	Re	48.164	**58.462**	49.084	0.879	47.736	6.891
	F1	58.085	**67.925**	55.273	1.667	36.964	6.128
Avg	Pr	77.237	**81.523**	75.272	56.219	58.052	47.762
	Re	78.494	**82.234**	77.544	20.203	52.601	55.991
	F1	77.089	**81.300**	75.302	21.157	51.211	45.521

Table 6. Experiment results for tweet humanitarian task for the CrisisBench dataset. Performance is reported in terms of Precision (Pr), Recall (Re) and F1-score (F1) for each class and also as a weighted average over all classes (Avg), with the best value of each metric displayed in **bold** within each row.

Class	Metric	Models					
					Llama 2		
		Bi-LSTM	BERT	CapsNet	0-shot	1-shot	5-shot
AI	Pr	75.118	80.910	**81.212**	5.478	10.612	17.035
	Re	76.975	**78.372**	66.281	94.653	14.017	3.569
	F1	75.838	**79.587**	72.931	10.357	11.057	5.231
CA	Pr	62.683	**70.034**	66.864	9.804	39.558	25.874
	Re	66.724	**70.783**	57.126	0.861	14.846	8.922
	F1	64.445	**70.349**	61.605	1.582	17.680	7.723
DE	Pr	50.211	**64.992**	29.580	2.098	15.696	2.631
	Re	38.384	**55.892**	7.744	9.091	30.640	54.545
	F1	42.958	**59.485**	12.272	3.409	19.930	4.958
DV	Pr	76.623	**76.708**	70.138	30.899	52.821	39.792
	Re	74.885	**82.048**	78.007	32.943	19.521	27.553
	F1	75.637	**79.254**	73.860	31.888	27.229	28.149
IUD	Pr	66.841	**73.555**	65.263	44.444	69.756	47.881
	Re	65.930	**67.864**	63.554	0.400	16.201	4.303
	F1	66.240	**70.518**	64.367	0.794	25.369	7.452
IDP	Pr	82.616	**82.860**	78.888	12.658	35.423	47.578
	Re	80.083	**84.551**	74.120	1.783	18.835	7.857
	F1	81.323	**83.572**	76.420	3.125	23.787	13.229
MFP	Pr	47.643	63.193	39.958	**100**	30.261	19.444
	Re	38.511	**39.482**	9.385	0.971	9.709	1.618
	F1	42.509	**48.595**	14.259	1.923	14.279	2.962
NH	Pr	90.170	**91.615**	87.765	90.020	68.583	71.694
	Re	93.639	**94.535**	94.258	26.641	85.396	67.854
	F1	91.865	**93.051**	90.892	41.115	75.426	64.895
RN	Pr	90.957	**94.112**	87.298	1.299	62.947	53.372
	Re	85.017	**90.414**	82.748	0.146	22.884	16.956
	F1	87.833	**92.220**	84.868	0.263	31.023	11.003
RE	Pr	36.993	**48.832**	13.889	0.000	7.958	10.038
	Re	16.139	**30.317**	0.754	0.000	13.122	4.676
	F1	21.325	**37.374**	1.431	0.000	9.373	6.107
SS	Pr	75.628	**82.361**	75.047	21.875	43.781	25.838
	Re	64.475	**69.085**	59.739	0.687	38.954	30.065
	F1	69.268	**75.140**	66.307	1.332	39.569	26.226
Avg	Pr	83.956	**86.679**	81.381	61.200	59.764	57.513
	Re	84.413	**87.000**	82.953	22.483	59.158	46.812
	F1	83.988	**86.708**	81.778	27.664	55.181	44.542

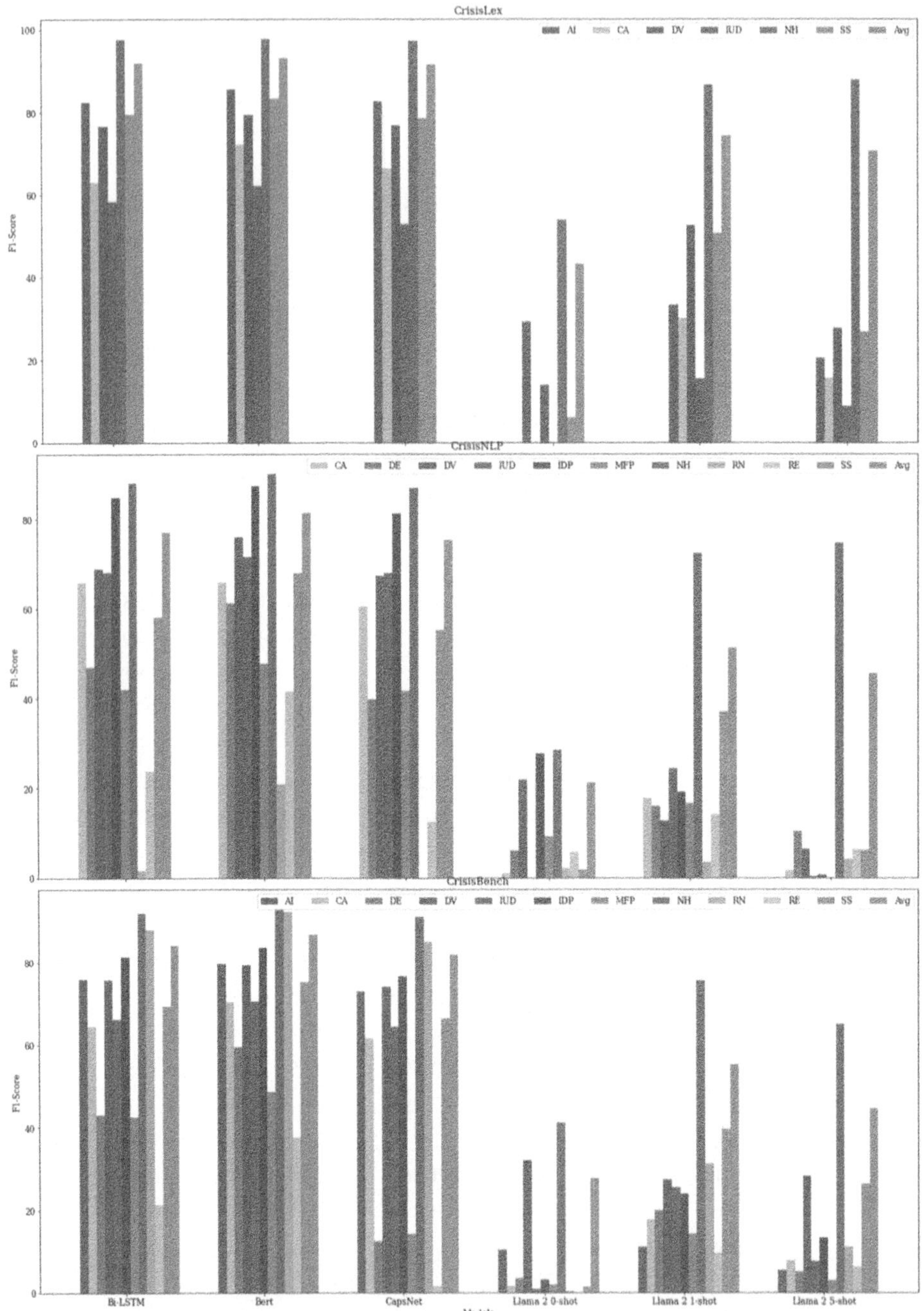

Fig. 2. The graph for F1-score comparison of Humanitarian task for the three datasets, CrisisNLP CrisisNLP, CrisisBench, respectively.

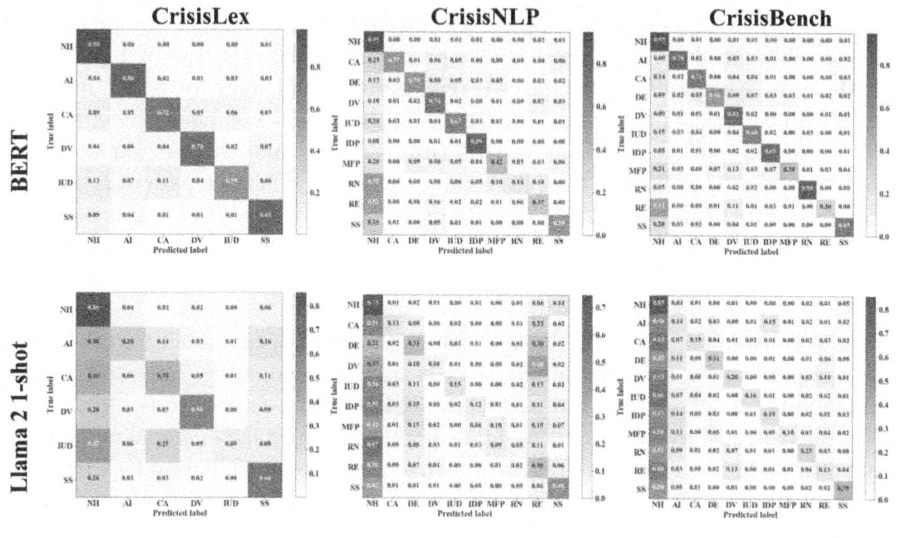

Fig. 3. Confusion matrix for humanitarian task for BERT and Llama 2 1-shot models for the three datasets.

for the Informative class, and 93.436, 83.993, and 84.862 for the Not-informative class in the CrisisLex, CrisisNLP, and CrisisBench datasets, respectively. Bi-LSTM and CapsNet models consistently secure the second and third positions, respectively, with comparable F1-scores. As for the low-resource model Llama 2, the ranking is 5-shot, 0-shot, and 1-shot for CrisisLex and CrisisBench datasets, while for CrisisNLP dataset, the ranking is 5-shot, 0-shot, and 1-shot for both classes and the average. The 1-shot variant consistently performs the worst in all cases for F1-score. This observation suggests that providing only one random example for each class, which may not be a representative sample, confuses the model more than helping it, compared to the scenario where no instances are shown to the model in 0-shot. On the other hand, with 5-shot, there is a greater chance for the model to "understand" commonalities among the five instances of each class, allowing it to generalize better to instances from the test dataset.

For the humanitarian categorization of crisis tweets, the F1-scores achieved by BERT models are 93.008, 81.300, and 86.708 for the CrisisLex, CrisisNLP, and CrisisBench datasets, respectively. Similar to the observed trend in the crisis tweet informativeness tasks across all three datasets, here, too, the F1-scores of BERT surpass those of the other five models. The ranking of the models, based on F1-scores for this classification task, is as follows for each dataset: Bi-LSTM, CapsNet, Llama 2 1-shot, 5-shot, and 0-shot. Specifically, for CrisisLex, the F1-scores are 91.850, 91.609, 74.167, 70.534, and 43.287 for Bi-LSTM, CapsNet, Llama 2 1-shot, 5-shot, and 0-shot, respectively. For CrisisNLP, the corresponding F1-scores are 77.089, 75.302, 51.211, 45.521, and 21.157. Lastly, for CrisisBench, the F1-scores are 83.988, 81.778, 44.542, 27.664, and 55.181. It is worth noting that BERT consistently outperforms the other models in terms of F1-scores across all classes and all three datasets.

BERT's classification of instances into different classes can be comprehensively examined through the confusion matrices presented in the first row of Fig. 3 for the three datasets. Analyzing these figures reveals certain challenges for BERT in specific class identifications. For the CrisisLex dataset, the model exhibits difficulties accurately identifying instances related to *IUD* (Infrastructure and Utilities Damage), with an accuracy of only 0.59. In CrisisNLP, a notable misclassification pattern is observed, where a majority of instances from *RN* (Requests or Needs) and *RE* (Response Efforts) classes, with accuracy values of 0.14 and 0.37, respectively, are misclassified as *NH* (Not Humanitarian). Similarly, in CrisisBench, a significant number of instances from the *RE* class, with an accuracy as low as 0.30, are misclassified as *NH*.

The confusion matrices for Llama 2 1-shot, the best Llama model for the humanitarian categories task, across different datasets are presented in Fig. 3, as well. A closer examination of these matrices reveals that the 1-shot setting can correctly identify some classes, as indicated by the presence of blue highlights along the diagonal of the confusion matrices. In contrast to the informativeness task, the humanitarian task involves a more fine-grained classification of informative instances, potentially explaining why 1-shot outperforms 0-shot in most cases. However, the introduction of additional random instances in the 5-shot setting seems to introduce some confusion to the model, resulting in a performance degradation compared to 1-shot. Nevertheless, the 5-shot setting consistently outperforms the 0-shots setting in this task.

Based on the results of both tasks, it can be concluded that BERT stands out as the most effective model for classifying crisis-related tweet datasets, demonstrating superior performance compared to the other models. The exceptional performance of BERT can be ascribed to various critical elements. Firstly, BERT's extensive pre-training on a large dataset allows it to grasp intricate patterns and contextual nuances deeply. Moreover, its incorporation of attention mechanisms and transformer architecture provides BERT with the capability to capture long-range dependencies in language, leading to a more comprehensive understanding and generation of natural language. Although CapsNet exhibits potential in certain image-related tasks due to its equivariant property preserving spatial relationships, it has not achieved a comparable level of success to BERT in the domain of natural language processing.

Moreover, Llama 2 models, in both zero- and few-shot settings, exhibit the least favorable performance, with a substantial performance gap. This underscores the importance of fine-tuning models with large corpus of datasets to enhance their effectiveness. It is worth acknowledging that despite the lack of fine-tuning and exposure to minimal to none instances in each class, Llama 2 models still perform reasonably well, highlighting their potential for tasks with limited labeled data, including crisis-related tweet classifications in the first hours of a disaster.

5 Related Work

Text classification, a prominent task in natural language processing (NLP), has been extensively studied across various application domains [26]. It has many applications, with one notable example being the filtering and classification of disaster-related messages on social media for effective disaster response.

Various traditional machine learning (ML) and natural language processing (NLP) techniques have been applied to filter disaster-related tweets [16]. However, since the manual extraction of features required by ML methods is time-consuming and may result in the loss of crucial textual elements, in recent years, there has been a shift towards using deep learning (DL) approaches for tweet filtering in disaster response [20,28,42]. For instance, Caragea et al. [6] utilized convolutional neural networks (CNNs) to identify informative tweets during disasters, while Neppalli et al. [36] conducted a comparative study between DL models (such as CNNs and recurrent neural networks) and traditional ML classifiers. The results showed that DL models, trained on a small amount of labeled data, outperformed standard ML classifiers, with the CNN model demonstrating superior performance over the RNN model. Additionally, Roy et al. [42] proposed a hybrid CNN model combining word and character-level embeddings for classifying tweets into various disaster-related categories. Kabir et al. [20] formulated a deep learning strategy by integrating attention-based Bi-LSTM networks with convolutional neural networks and constructing an auxiliary feature map for tweet categorization.

Most recently, several investigations have delved into applying transformer-based models for the classification of disaster-related tweets [7,13,33,35,38,41,53]. For instance, Ma et al. [32] demonstrated that the BERT model exhibited superior performance compared to the Bi-LSTM model but showed similar performance to customized BERT-based models that integrated BERT with LSTM or CNN networks (i.e., BERT+LSTM and BERT+CNN). In another study, Alam et al. [5] conducted a comparative analysis of three transformer-based models (BERT, DistilBERT, RoBERTa) against CNN and FastText models for both informativeness and humanitarian tasks using three datasets, including the CrisisBench benchmark dataset. The findings revealed that transformer-based models surpassed CNN and FastText models, with comparable performance among the transformer-based models. Based on these outcomes from previous research, we acknowledge BERT as a state-of-the-art model for disaster tweet classification, and thus, we incorporate it as a strong model in our study.

In recent years, the CapsNet model, introduced by Sabour et al. [43], has gained significant attention in the field of image classification. Leveraging the dynamic routing mechanism, CapsNets excel at preserving spatial relationships between features, addressing a fundamental constraint of CNNs related to information loss due to pooling for achieving translation invariance [40]. While widely applied across various domains, CapsNet models have also shown promise in the realm of disaster image classification [12]. Building on the success observed in image classification, Zhao et al. [56] investigated the use of CapsNets with dynamic routing for text classification, demonstrating their capability to achieve competitive results.

Following the work of Zhao et al. [56], numerous researchers have employed the CapsNet model or its variations for text classification across diverse application domains [9,23,24,50,51,54–56]. For instance, Yang et al. [51] introduced a cross-domain capsule network (TL-Capsule) showcasing its knowledge transfer capabilities in text classification. Meanwhile, Kim et al. [24] enhanced CapsNets by incorporating an ELU-gate and introducing a static routing procedure, with a comparative study on seven benchmark datasets revealing the superiority of the static routing variant over dynamic routing. Additionally, Demotte et al. [9] explored shallow and deep CapsNets with both dynamic and static routing for social media content analysis, finding that shallow CapsNets with static routing outperformed their deep counterparts with either dynamic or static routing.

Despite the use of CapsNets for text classification and its proven superiority over CNN and RNN models, to the best of our knowledge, there is no prior study that extensively compares CapsNets with BERT-type models on text classification. Furthermore, CapsNets have not been explored for disaster tweet classification, although they have shown promising results on disaster image classification [12]. Therefore, the objective of our work was to perform a comprehensive comparison between CapsNet and BERT models on disaster tweet classification. We focused on CapsNets with static routing [24] and addressed two specific tasks useful for disaster response: filtering tweets based on informativeness (a binary classification task) and identifying a tweet's humanitarian category (a multi-class classification task). We used three existing datasets in our experiments: CrisisLex, CrisisNLP, and CrisisBench, a benchmark dataset consolidating eight existing datasets [5]. In addition to the CapsNet and BERT models, we include Bi-LSTM models in our study given that these models have been used as strong baselines for CapsNet and BERT models in prior works [50,54–56].

Furthermore, significant advancements in large language models (LLMs) underscore their proficiency in complex reasoning tasks across diverse domains [37]. Notably, the Llama 2 model has demonstrated competitiveness with its extensive pretraining and substantial parameter size, showcasing remarkable generalization capacities in zero- and few-shot learning [46]. This makes Llama 2 an excellent choice for our application domain, particularly when dealing with scenarios where few or no labels are provided during the initial hours of a disaster. Consequently, we have incorporated experiments with the Llama 2 model in both zero- and few-shot settings.

6 Conclusions and Future Work

In this study, we conducted an in-depth evaluation of fine-tuned CapsNets, BERT, and Bi-LSTM models by comparison with Llama 2 in 0-shot and few-shot settings, for crisis-related tweet classification tasks, specifically focusing on tweet informativeness (binary classification) and tweet humanitarian categorization (multi-class classification). The evaluation utilized three datasets: CrisisLex, CrisisNLP, and CrisisBench. The findings underscored that among fine-tuned deep learning models, BERT consistently demonstrated the best performance in classifying crisis-related data, while Bi-LSTM and CapsNet models exhibited comparable results. For Llama 2 models in 0-shot and few-shot settings, performance was not directly comparable with that of the fine-tuned models. However, considering that no training instances (0-shot) or only a few instances (few-shot) were presented to the Llama 2 model as part of the prompts, the results were noteworthy. The F1-score results indicate that 0-shot settings yielded best results for informativeness tasks, while the 1-shot setting demonstrated best overall performance for humanitarian tasks. This trend suggests a notable influence of prompt selection and instance choice, emphasizing the importance of a careful selection process for instances in few-shot scenarios to showcase representative examples for each class, avoiding random selection.

Our future work aims to explore the potential of more recent transformer-based models for disaster tweet classification considering the superiority of BERT models over CapsNet and Bi-LSTM models. Additionally, we plan to include the application of these models to a dataset containing both textual and visual modalities, aiming to develop a multi-modal model based on transformer architectures for classifying disaster-related posts. Recognizing the complementary nature of textual and visual information, leveraging both modalities simultaneously is anticipated to lead to enhanced performance. Furthermore, given the dependency of Llama 2 on prompts and example selection, our future research will delve into alternative approaches for creating optimal prompts for large language models, including Llama 2, in zero and few-shot settings. This exploration will involve a comparative analysis with fine-tuned models in the context of crisis-related content classification.

Acknowledgements. We thank the National Science Foundation and Amazon Web Services for support from grant IIS-1741345, which supported the research and the computation in this study.

References

1. Acheampong, F.A., Nunoo-Mensah, H., Chen, W.: Transformer models for text-based emotion detection: a review of BERT-based approaches. Artif. Intell. Rev. **54**(8), 5789–5829 (2021)
2. Ahmed, A.: Use of social media in disaster management (2011)
3. Alam, F., Alam, T., Hasan, M.A., Hasnat, A., Imran, M., Ofli, F.: Medic: a multi-task learning dataset for disaster image classification. Neural Comput. Appl. **35**(3), 2609–2632 (2023)
4. Alam, F., Ofli, F., Imran, M., Aupetit, M.: A Twitter tale of three hurricanes: Harvey, Irma, and Maria. arXiv preprint arXiv:1805.05144 (2018)
5. Alam, F., Sajjad, H., Imran, M., Ofli, F.: Crisisbench: benchmarking crisis-related social media datasets for humanitarian information processing. arXiv preprint arXiv:2004.06774 (2020)
6. Caragea, C., Silvescu, A., Tapia, A.H.: Identifying informative messages in disaster events using convolutional neural networks. In: International Conference on Information Systems for Crisis Response and Management, pp. 137–147 (2016)
7. Chanda, A.K.: Efficacy of BERT embeddings on predicting disaster from Twitter data. arXiv preprint arXiv:2108.10698 (2021)
8. Chen, Z., Li, S., Ye, L., Zhang, H.: Multi-label classification of legal text based on label embedding and capsule network. Appl. Intell. **53**(6), 6873–6886 (2023)
9. Demotte, P., Wijegunarathna, K., Meedeniya, D., Perera, I.: Enhanced sentiment extraction architecture for social media content analysis using capsule networks. Multimed. Tools Appl. 1–26 (2021)
10. Devlin, J., Chang, M.W., Lee, K., Toutanova, K.: BERT: pre-training of deep bidirectional transformers for language understanding. arXiv preprint arXiv:1810.04805 (2018)
11. Dinani, S., Caragea, D.: A comparison study for disaster tweet classification using deep learning models. In: Proceedings of the 12th International Conference on Data Science, Technology and Applications DATA, vol. 1. SCITEPRESS-Science and Technology Publications (2023)

12. Dinani, S.T., Caragea, D.: Disaster image classification using capsule networks. In: 2021 International Joint Conference on Neural Networks (IJCNN), pp. 1–8. IEEE (2021)
13. Dinani, S.T., Caragea, D.: Disaster image classification using pre-trained transformer and contrastive learning models. In: 2023 IEEE 10th International Conference on Data Science and Advanced Analytics (DSAA), pp. 1–11. IEEE (2023)
14. Graves, A., Schmidhuber, J.: Framewise phoneme classification with bidirectional LSTM and other neural network architectures. Neural Netw. **18**(5–6), 602–610 (2005)
15. Hinton, G.E., Sabour, S., Frosst, N.: Matrix capsules with EM routing. In: International Conference on Learning Representations (2018)
16. Imran, M., Castillo, C., Diaz, F., Vieweg, S.: Processing social media messages in mass emergency: survey summary. In: Companion Proceedings of the Web Conference 2018, pp. 507–511 (2018)
17. Imran, M., Mitra, P., Castillo, C.: Twitter as a lifeline: human-annotated twitter corpora for NLP of crisis-related messages. arXiv preprint arXiv:1605.05894 (2016)
18. Imran, M., Ofli, F., Caragea, D., Torralba, A.: Using AI and social media multimodal content for disaster response and management: opportunities, challenges, and future directions (2020)
19. Joulin, A., Grave, E., Bojanowski, P., Mikolov, T.: Bag of tricks for efficient text classification. arXiv preprint arXiv:1607.01759 (2016)
20. Kabir, M.Y., Madria, S.: A deep learning approach for tweet classification and rescue scheduling for effective disaster management. In: Proceedings of the 27th ACM SIGSPATIAL International Conference on Advances in Geographic Information Systems, pp. 269–278 (2019)
21. Kaliyar, R.K.: A multi-layer bidirectional transformer encoder for pre-trained word embedding: a survey of BERT. In: 2020 10th International Conference on Cloud Computing, Data Science & Engineering (Confluence), pp. 336–340. IEEE (2020)
22. Khajwal, A.B., Cheng, C.S., Noshadravan, A.: Post-disaster damage classification based on deep multi-view image fusion. Comput.-Aided Civil Infrastruct. Eng. **38**(4), 528–544 (2023)
23. Khikmatullaev, A.: Capsule Neural Networks for Text Classification. Universitat Bonn (2019)
24. Kim, J., Jang, S., Park, E., Choi, S.: Text classification using capsules. Neurocomputing **376**, 214–221 (2020)
25. Koren, M.: Using Twitter to save a newborn from a flood. The Atlantic (2017). https://www.theatlantic.com/technology/archive/2017/08/harvey-rescue-twitter/538191/
26. Kowsari, K., Jafari Meimandi, K., Heidarysafa, M., Mendu, S., Barnes, L., Brown, D.: Text classification algorithms: a survey. Information **10**(4), 150 (2019)
27. Li, H., Caragea, D., Caragea, C., Herndon, N.: Disaster response aided by tweet classification with a domain adaptation approach. J. Contingencies Crisis Manag. **26**(1), 16–27 (2018)
28. Li, X., Caragea, D.: Domain adaptation with reconstruction for disaster tweet classification. In: Proceedings of the 43rd International ACM SIGIR Conference on Research and Development in Information Retrieval, pp. 1561–1564 (2020)
29. Liu, G., Guo, J.: Bidirectional LSTM with attention mechanism and convolutional layer for text classification. Neurocomputing **337**, 325–338 (2019)
30. Liu, Y., et al.: RoBERTa: a robustly optimized BERT pretraining approach. arXiv preprint arXiv:1907.11692 (2019)
31. Lu, G., Gan, J., Yin, J., Luo, Z., Li, B., Zhao, X.: Multi-task learning using a hybrid representation for text classification. Neural Comput. Appl. **32**(11), 6467–6480 (2020)
32. Ma, G.: Tweets classification with BERT in the field of disaster management. StudentReport@ Stanford. edu (2019)

33. Maharani, W.: Sentiment analysis during Jakarta flood for emergency responses and situational awareness in disaster management using BERT. In: 2020 8th International Conference on Information and Communication Technology (ICoICT), pp. 1–5. IEEE (2020)
34. Mazloom, R., Li, H., Caragea, D., Caragea, C., Imran, M.: A hybrid domain adaptation approach for identifying crisis-relevant tweets. Int. J. Inf. Syst. Crisis Response Manag. (IJISCRAM) **11**(2), 1–19 (2019)
35. Naaz, S., Abedin, Z.U., Rizvi, D.R.: Sequence classification of tweets with transfer learning via BERT in the field of disaster management. EAI Endorsed Trans. Scalable Inf. Syst. **8**(31), e8 (2021)
36. Neppalli, V.K., Caragea, C., Caragea, D.: Deep neural networks versus Naive Bayes classifiers for identifying informative tweets during disasters. In: Proceedings of the 15th Annual Conference for Information Systems for Crisis Response and Management (ISCRAM) (2018)
37. Nguyen, T.T., Wilson, C., Dalins, J.: Fine-tuning llama 2 large language models for detecting online sexual predatory chats and abusive texts. arXiv preprint arXiv:2308.14683 (2023)
38. Ningsih, A., Hadiana, A.: Disaster tweets classification in disaster response using bidirectional encoder representations from transformer (BERT). In: IOP Conference Series: Materials Science and Engineering, vol. 1115, p. 012032. IOP Publishing (2021)
39. Olteanu, A., Castillo, C., Diaz, F., Vieweg, S.: Crisislex: a lexicon for collecting and filtering microblogged communications in crises. In: Eighth International AAAI Conference on Weblogs and Social Media (2014)
40. Patrick, M.K., Adekoya, A.F., Mighty, A.A., Edward, B.Y.: Capsule networks-a survey. J. King Saud Univ. Comput. Inf. Sci. **34**(1), 1295–1310 (2022)
41. Ray Chowdhury, J., Caragea, C., Caragea, D.: Cross-lingual disaster-related multi-label tweet classification with manifold mixup. In: Proceedings of the 58th Annual Meeting of the Association for Computational Linguistics: Student Research Workshop (2020)
42. Roy, P.K., Kumar, A., Singh, J.P., Dwivedi, Y.K., Rana, N.P., Raman, R.: Disaster related social media content processing for sustainable cities. Sustain. Urban Areas **75**, 103363 (2021)
43. Sabour, S., Frosst, N., Hinton, G.E.: Dynamic routing between capsules. In: Advances in Neural Information Processing Systems, vol. 30 (2017)
44. Sanh, V., Debut, L., Chaumond, J., Wolf, T.: DistilBERT, a distilled version of BERT: smaller, faster, cheaper and lighter. arXiv preprint arXiv:1910.01108 (2019)
45. Sathishkumar, V.E., Cho, J., Subramanian, M., Naren, O.S.: Forest fire and smoke detection using deep learning-based learning without forgetting. Fire Ecol. **19**(1), 1–17 (2023)
46. Touvron, H., et al.: LLaMA: open and efficient foundation language models. arXiv preprint arXiv:2302.13971 (2023)
47. Touvron, H., et al.: LLaMA 2: open foundation and fine-tuned chat models. arXiv preprint arXiv:2307.09288 (2023)
48. Vaswani, A., et al.: Attention is all you need. In: Advances in Neural Information Processing Systems, vol. 30 (2017)
49. Villegas, C., Martinez, M., Krause, M.: Lessons from Harvey: crisis informatics for urban resilience. Rice University Kinder Institute for Urban Research (2018)
50. Wu, Y., Li, J., Wu, J., Chang, J.: Siamese capsule networks with global and local features for text classification. Neurocomputing **390**, 88–98 (2020)
51. Yang, M., Zhao, W., Chen, L., Qu, Q., Zhao, Z., Shen, Y.: Investigating the transferring capability of capsule networks for text classification. Neural Netw. **118**, 247–261 (2019)
52. Young, C.E., Young, C.E., Kuligowski, E.D., Pradhan, A.: A review of social media use during disaster response and recovery phases. US Department of Commerce, National Institute of Standards and Technology (2020)

53. Zahera, H.M., Elgendy, I.A., Jalota, R., Sherif, M.A.: Fine-tuned BERT model for multi-label tweets classification. In: TREC, pp. 1–7 (2019)
54. Zhang, B., Xu, X., Yang, M., Chen, X., Ye, Y.: Cross-domain sentiment classification by capsule network with semantic rules. IEEE Access **6**, 58284–58294 (2018)
55. Zhao, W., Peng, H., Eger, S., Cambria, E., Yang, M.: Towards scalable and reliable capsule networks for challenging NLP applications. arXiv preprint arXiv:1906.02829 (2019)
56. Zhao, W., Ye, J., Yang, M., Lei, Z., Zhang, S., Zhao, Z.: Investigating capsule networks with dynamic routing for text classification. arXiv preprint arXiv:1804.00538 (2018)

Statistical Hypothesis Testing and Modelling of Peoples' Power: A Causal Study of the #BlackLivesMatter Movement via Hawkes Processes on Social and Mass Media

Alfred Lindström[1,2(✉)], Simon Lindgren[3], and Raazesh Sainudiin[1,2]

[1] Department of Mathematics, Uppsala University, Uppsala, Sweden
raazesh.sainudiin@math.uu.se
[2] Combient Competence Centre for Data Engineering Sciences, Uppsala University, Uppsala, Sweden
lindstrom.alfred@gmail.com
[3] Department of Sociology, Umeå University, Umeå, Sweden
simon.lindgren@umu.se

Abstract. In the current mass media landscape with a few corporate owners and operating under the propaganda model of communication aimed at manufacturing system-supportive consent, and the algorithmic-rent seeking business models of most popular social media platforms, we set out to ask whether Peoples still have power to take collective real-world action that may be counter to prevailing media tendencies. We study interactions in social media and the reports in mass media during the Black Lives Matter (BLM) protests following the death of George Floyd. We implement open-source pipelines to process the data at scale and employ the self-exciting counting process known as Hawkes process to address our main question: is there a causal relation between interactions in social media and reports of street protests in mass media? Specifically, we use network models to identify such interactions in Twitter, that supported the BLM movement, and compared the timing of these interaction to those of news reports of street protests mentioning George Floyd, via the Global Database of Events, Language, and Tone (GDELT) Project. The comparison was made through a Bivariate Hawkes process model for a formal hypothesis test of Granger-causality. We show that interactions in social media that supported the BLM movement, at the beginning of nationwide protests, caused the global mass media reports of street protests in solidarity with the movement. We also use more general Hawkes process model to understand the diffusion of specific influential messages in social media. Our study suggests that BLM activists have harnessed social media to mobilise street protests across the planet despite the concentrated ownership of mass media and the algorithmic rent-seeking business models of social media platforms.

Keywords: Hawkes process · Community detection · Granger causality · Hypothesis test · Social · Mass media modelling

Swedish Research Council project 2019-03351 & the Wallenberg AI, Autonomous Systems & Software Program funded by Knut and Alice Wallenberg Foundation.

O. Gusikhin et al. (Eds.): DATA 2023, CCIS 2105, pp. 95–126, 2024.
https://doi.org/10.1007/978-3-031-68919-2_5

1 Introduction

Although social media, at least in many nation states, can generally be used by any individual or group to spread their messages on any topic, mass media ownership is highly concentrated in the hands of a few corporate actors in several nation states under different governance models, including those of the USA [1] and member states of the European Union [39]. On the other hand, both social and mass media are largely "harmonised" by the state in People's Republic of China [38]. It is well-established in "Manufacturing Consent: The Political Economy of the Mass Media" [20] by Edward S. Herman and Noam Chomsky that the mass communication media of the USA "are effective and powerful ideological institutions that carry out a system-supportive propaganda function, by reliance on market forces, internalised assumptions, and self-censorship, and without overt coercion", by means of the so-called *propaganda model of communication*, whereby "the manufacture of consent" refers to consent of the governed, and derives from the phrase used in 1922 by Walter Lippmann [25].

Most popular social media platforms are owned by corporations and are not bastions of individual free expressions. Their business models have been articulated more generally as one of *surveillance capitalism* by Zuboff [43] and specifically how social media disrupt our elections, our economy and our health has been expounded by Aral [2], while Frenkel and Kang [14] focus on a specific social media platform, to mention a few recent studies. A recent theory of *algorithmic attention rents* [31] in digital aggregator platforms, including social media, explores how such platforms become increasingly capable of extracting rents from a variety of actors in their ecosystems – users, suppliers, and advertisers – through their algorithmic control over user attention, and how regulations mandating the disclosure of their operating metrics and details on how user attention is monetised are urgently needed.

Given the concentration of mass media ownership by a few corporate actors in many states in the West and their propaganda model of communication that routinely manufactures consent, and the relative freedom of individual communications of free expressions in social media which aims to monetise user attention, we set out to conduct a detailed statistical analysis through mathematical modelling of interactions between Twitter, the social media platform, and mass media, surrounding the #BlackLivesMatter movement during the summer of 2020 to ask the most basic question: *Do people still have power to mobilise and take collective real-world actions?*. In short, we find in this study that the answer to this question, albeit limited to a specific formulation, is a definite yes.

1.1 BLM-Movement

On 25th of May 2020, George Floyd, a 46 year old African-American man, is arrested in Minneapolis, Minnesota for allegedly using a counterfeit $20 bill to buy cigarettes. The arrest is caught on film by passersby, showing how police officer Derek Chauvin pins the handcuffed Floyd to the ground with his knee on Floyd's neck, while his three colleagues prevent anyone from intervening. Floyd repeatedly utters the words "I can't breathe" before he goes unconscious. He later dies at the hospital, and the video of the arrest goes viral on Facebook [10]. The next day protests in support of the Black

Lives Matter (BLM) movement, and against police brutality, start in Minneapolis, which during the following days will spread both nationally and internationally to over 60 countries, and become what may be the largest protests in U.S. history to date, with polls estimating attendances in the range of 15–26 million people [8].

BLM is a decentralised grassroots movement that began on social media, using the hashtag #BlackLivesMatter in the wake of the shooting of Trayvon Martin in July 2013. The movement has since then gained attention for demonstrations following the deaths of Michael Brown and Eric Garner in 2014, and George Floyd in 2020, with its main issues being that of advocating against police brutality toward African-Americans, and policy issues related to racial injustices [21].

As reactions and critiques of the BLM movement, the phrase "All lives matter" was coined, as well as the phrase "Blue lives matter", after the shooting of two police officers during protests in Ferguson, Missouri in 2015. Both of these slogans are associated with conservative views, and rejects the BLM-movement's idea of a need to focus on the racial injustice towards African Americans.

The decentralised nature of all three of these movements, and the way social media has played a key part in their development, leading to real life events such as mass protests, motivates our choice to analyse data from social media and from mass media to try to get a better understanding of the mobilisation in social media into real-world action.

1.2 Outline

In this work we study the landscape in mass and social media during the first month of protests that followed after the murder of George Floyd. Our primary question is whether there is a statistically significant causal interaction between communications in socially networked communities and street protests as measured by published reports in mass media. We attempt to answer this question by devising a data processing framework to mathematically model the interactions between social and mass media via the family of point processes known as Hawkes processes and conduct statistical hypothesis tests of Granger causality, subsequent to identifying influential social media communities, often engaged in ideological competition, using network models. Furthermore, we focus on the social media diffusion process towards understanding how and when information is spread influentially as cascades into the socially networked communities.

Briefly, this Chapter's outline is as follows. We describe Models in Sect. 2, Data Handling in Sect. 3, Analysis of Twitter Data in Sect. 4, Joint Media Modelling in Sect. 5, Modelling Retweet Cascades in Sect. 6, and conclude in Sect. 7.

1.3 Extending DATA2023 Conference Paper

This Chapter is a significant extension of a conference paper by the same authors [24]. Section 6 is entirely new while Sect. 2 has been expanded to include more general models and make the mathematical exposition more self-contained. The other Sections are largely excerpted from the conference paper except Sect. 5 with additional details on stabilising nonlinear optimisation. Section 1 here frames a much broader context and Sect. 7 is also appropriately extended.

2 Models

2.1 Hawkes Processes

We will now introduce a family of point processes known as Hawkes processes, assuming the reader is familiar with point processes. These processes were introduced by Hawkes [19], and due to their self-exciting nature they are used in fields such as epidemiology, seismology, and finance [5,9]. Moreover, it has been implemented in analysis of diffusion processes in social media [28,41].

Suppose we observe events in continuous time, i.e., points on the positive real line as *timestamps*, where for each i, t_i is the exact time where some sort of event occurs for the i-th time. Define the *history* of a point process up to time t, as the set $\mathcal{H}_t$ containing all timestamps $\{t_i\}$ up to time t. A Hawkes process allows us to model the occurrence of future events after time t based on the entire history $\mathcal{H}_t$ up to time t as follows:

Definition 1. *Let $N(t)$ be a point process that counts the number of events up to time t with history $\mathcal{H}_t$. If the intensity $\lambda(t)$ of $N(t)$ is of the form*

$$\lambda(t) = \mu + \sum_{t_i \in \mathcal{H}_t} \phi(t - t_i), \tag{1}$$

we define $N(t)$ as a Hawkes process, where μ is the baseline intensity and $\phi(t)$ is the kernel (Eq. (1) in [24]).

The events (i.e., points on the positive real line) of a Hawkes process, can be interpreted as being of two types. First we have the *immigrants* which arrive at a constant rate of the baseline intensity μ. Next, we have the *offspring* which are produced by existing events. These arrive after time t via the intensity of the kernel ϕ, from any historical event $t_i \in \mathcal{H}_t$, which is often chosen to be monotonically decreasing, and is thus a descendant of an already existing event in history. Note that all events in history, whether they are immigrants or offspring, may produce new offspring.

As the arrival of an event increases the rate of new events arriving close in time, intuitively we can talk about Hawkes processes having a self-exciting nature; events will naturally cluster around immigrant events. For a concrete example, one can think of the immigrant events as an earthquake occurring, with the offspring being after-shocks.

For a Hawkes process to be stationary, we require some constraints on the kernel.

Definition 2. *Let $\phi(t)$ be a kernel for a Hawkes process N. We define*

$$\nu = \int_0^\infty \phi(t)dt, \tag{2}$$

as the branching factor of the Hawkes process.

The branching factor tells us, the mean number of offspring events one event can have. If $\nu < 1$, the process is in the *subcritical region*, and the branching from one event will die out. If $\nu > 1$, it is in the *supercritical region* and will explode exponentially. Moreover, if $\nu < 1$ then inductively we get the estimate, via the geometric sum, that an event will generate $1/(1 - \nu)$ offspring in total on average.

One instructive example of how to interpret the branching factor comes from Filimonov and Sornette [13], where they examine trading by looking at financial data. Their estimation of the branching factor $\nu \in (0.7, 0.8)$, means that 70%-80% of all the trades in the given data are due to past trades, rather than external events happening.

Theorem 1. *A Hawkes process is stationary if and only if its branching factor satisfies* $\nu < 1$.

We will now introduce a particular choice of kernel.

Definition 3. *We define*

$$\phi(t) = \alpha\beta e^{-\beta t}, \tag{3}$$

as an exponential kernel where parameter $\alpha \geq 0$ *is the self-excitation parameter, and parameter* $\beta > 0$ *is the decay rate (Eq. (2) in [24]).*

Parameter α thus decides how much an occurred event will influence the rate of new events, while β will decide how long into the future this influence will last as $\phi(t) \to 0$, when $t \to \infty$.

Marked Hawkes Point Processes. We now introduce an extension of Hawkes processes, where each timestamp of an event not only contains the event's location in time, but also some information about that specific event. These types of point processes are known as marked point processes. Each mark or type is represented by an $i \in \{1, 2, \ldots, d\}$, the set of possible marks that is seen to represent one of d dimensions.

Definition 4. *Let* N^* *be a point process on* $\mathbb{R}_+$*. We define* N^* *as the ground process. Let the mark space* $\mathcal{K}$ *be a set of marks. We then define the point process* $\{t_i, \kappa_i\}$ *on* $\mathbb{R}_+ \times \mathcal{K}$ *as a marked point process, if* N^* *is a well-defined point process, i.e., for bounded* $A \in \mathbb{R}_+$, $N_g(A) = N(A \times \mathcal{K}) < \infty$.

An example of a marked point process would be a model where we in one dimension have the timestamps of earthquakes occurring, and in the mark space we have the information of the magnitude of each earthquake. The size of magnitude could then influence the intensity of new earthquakes occurring in time. Other than the exponential kernel, We experimented with several other kernels, including the In Sect. 6, we will use a marked Hawkes process with a power-law kernel defined as bellow.

Definition 5. *Let* N *be marked point process with history* $\mathcal{H}_{t,m}$ *with elements on the form* $(t_i, m_i) \in \mathbb{R}_+ \times \mathbb{N}$*. If the intensity of* N *is of the form:*

$$\begin{aligned} \lambda(t) &= \sum_{(t_i,m_i)\in\mathcal{H}_{t,m}} \phi_{m_i}(t - t_i) \\ &= \sum_{(t_i,m_i)\in\mathcal{H}_{t,m}} \kappa m_i^{\beta}(t - t_i + c)^{-(1+\theta)}, \end{aligned} \tag{4}$$

where, $\kappa, \beta, \theta, c > 0$*, then* N *is a* marked Hawkes process *with* a mark-proportioned power-law kernel.

Analogous to the exponential kernel, $1 + \theta$ captures how quickly an event is forgotten, while parameter c shifts the term so that $\phi_{m_i}(t)$ is bounded when $t \approx 0$. Parameters κ and β determine how much an event influences the intensity of future events. Note that that this influence is affected by the mark $m_i \in \mathbb{N}$, i.e., the larger m_i is, the larger the influence of that event will be.

Multivariate Hawkes Processes. Next, we introduce a natural extension of Hawkes processes, where each timestamp of an event not only contains the event's location in time, but also some information about that specific event. Such point processes are known as marked point processes. Each mark or type is represented by a dimension $i \in \{1, 2, \ldots, d\}$, the set of possible marks It is natural to model Hawkes processes in multiple dimensions, where each dimension is a Hawkes process for each or types of events, that are possibly interconnected in terms of their influence or lack thereof on each other's occurrence. A natural extension of the Hawkes process is the multivariate Hawkes process. To continue with the example of earthquakes, let the first dimension in our multivariate Hawkes process count the number of earthquakes in a region, and the second dimension count the number of tsunamis in the same region. The arrival of an earthquake might then give rise to a tsunami, which a multivariate Hawkes process, given the right parameters, can capture.

Definition 6. *We define a marked point process N with mark space $\{1, \ldots, d\}$ as multivariate point process. Moreover, for $i \in \{1, \ldots, d\}$ we denote the i:th component of the multivariate point process N as N_i, where $N = (N_1, \ldots, N_d)$ is a marked point process, and N_i is a well-defined ground process. We refer to d as the dimension of N.*

Definition 7. *Let $d \in \mathbb{N}$ be the number of dimensions, and $\mathcal{H}_{t,i}$ for $i = 1, \ldots, d$ be the history of events in dimension i. The multivariate point-process induced by the intensities*

$$\lambda_i(t) = \mu_i + \sum_{j=1}^{d} \sum_{t_k \in \mathcal{H}_{t,j}} \phi_{ij}(t - t_k) \quad i = 1, ..., d \tag{5}$$

is then defined as a multivariate Hawkes process (Eq. (3) in [24]).

Definition 8. *Let $d \in \mathbb{N}$ be the number of dimensions. We If the kernel $\phi_{ij}(t)$ takes the form of the following multivariate exponential kernel,*

$$\phi_{ij}(t) = \alpha_{ij} \beta_{ij} e^{-\beta_{ij} t} \quad i, j = 1, ..., d, \tag{6}$$

where $\alpha_{ij} \geq 0$ is the excitation parameter, and $\beta_{ij} > 0$ is the decay rate, then we have the multivariate Hawkes process with exponential kernel (Eq. (4) in [24]).

The excitation parameter α_{ij} can be interpreted similarly as α in the one-dimensional case with the exponential kernel, with the exception that this influence on new events in dimension i now may come from previous events in any dimension $j \in \{1, \ldots, d\}$. Analogously, β_{ij} is interpreted as the rate of decay that specifies how past events in dimension j can influence the arrival of new events in dimension i. In Sect. 5 we use a multivariate Hawkes process to model Twitter events in dimension 1 and mass media reports of protests in dimension 2.

2.2 Granger Causality

How to rigorously define causality has been a topic of discussion in western philosophy for over 2000 years, starting with Plato and Aristotle [12], and continuing on with Hume and Kant's disagreement being one of the fundamental discussions in modern philosophy. The problem is still open, [32].

In light of this, and in some sense to get around the metaphysical complications of proper causality, Clive Granger introduced the concept of *Granger Causality* relating to stochastic processes. The basic idea is if a variable X_t *Granger-causes* variable Y_t, then the past values of X_t contain information that helps predict future values of Y_{t+1} better than doing prediction based only on past values of Y_t [18].

Using the following Theorem from Eichler [11], we will test the null hypothesis of the non-existence of Granger causality between events in social and mass media, and vice versa, in the sequel.

Theorem 2. *Let $N(t)$ be a multivariate Hawkes process in d dimensions, with kernels $\phi_{ij}(t)$, $i, j \in \{1, \ldots, d\}$. Then the j-th component N_j does not Granger-cause the i-th component N_i if and only if $\phi_{ij} = 0$, $\forall t \in \mathbb{R}$.*

Thus, when $N(t)$ is a multivariate Hawkes process with exponential kernel, by Theorem 2 the j-th component N_j does not Granger-cause the i-th component N_i if and only if $\alpha_{ij} = 0$, $\forall t \in \mathbb{R}$.

When examining Granger causality on more than two dimensions, it is natural to look at the following induced graph.

Definition 9. *Let N be a multivariate Hawkes process in d dimensions. We define the Granger causality graph G_c with vertices $V = \{1, \ldots, d\}$, directed edges $(u, v) \in E$ from u to v, and the following constraint*

$$(i, j) \notin E \Leftrightarrow \phi_{ji}(t) = 0, \quad \forall t, \text{ and } i, j \in V. \tag{7}$$

Via the Granger causality graph, one can naturally talk about indirect Granger causality; assume that there is no edge from vertices i to j, i.e., the i:th component does not Granger-cause the j:th component. The i:th component may however affect the j:th indirectly, if there exists a path from i to j in the Granger causality graph.

3 Data Handling

3.1 Apache SPARK

The data was handled using Apache Spark[1] which is an open-source engine designed for data engineering, data science, and machine learning on clusters of multiple computers, by implicit data parallelism. Spark is multi-language and supports `Scala`, `Python`, `R`, `SQL`, `Java`, `C#` and `F#`. While most of the code for this article was written in `Scala`, the ease of switching between languages in the same environment proved quite useful, as we would use libraries written in both `R` and `Python`.

[1] https://github.com/apache/spark.

On top of Spark core, Spark SQL [4], which introduces the data abstraction of DataFrames, allows manipulation in `Scala`, `Python`, and `R` using the standard `SQL` language, and the graph-processing framework GraphX [17], allows for network-analysis. To run Spark, the cloud data platform Databricks was used, which provided cloud storage, computing clusters, and a notebook-environment to write and run the code after loading the two main libraries developed for this study, `MEP`[2] and `SPARK-GDELT`[3].

3.2 Twitter

Twitter is a micro-blog and social media service, founded in 2006, where users post and interact via *tweets* – a short message restricted to 280 characters, which may also contain pictures, short videos and URLs. Tweets can be original posts, replies to other tweets, or retweets, i.e., sharing of another user's tweet. As long as a user does not actively chose to be private, anyone is able to read the tweets of the user. To help a tweet gain attraction, and make it easier for other users to find tweets on a specific topic, the user can tag their posts by including keywords prefaced with '#', the hash symbol. These tagged keywords are called *hashtags* and they have been used by activists in global social movements such as #BlackLivesMatter and #MeToo to raise awareness of injustice and counter prevailing narratives [21].

Users may also follow other users on Twitter. The relationship of following is asymmetrical, meaning that if user A follows user B, user B does not have to follow user A. Compare this to Facebook, where users mutually have to accept each other as friends to be able to interact. To simplify things, if Facebook is about keeping in touch and networking with your friends, Twitter is about sharing and receiving information the user finds interesting; according to a study done in 2014, 44% of Twitter's users have never tweeted which seems to suggest that a large part of the user base only uses Twitter for receiving information [29]. As of the fourth quarter of 2020, Twitter has 192 million daily users [40]. Due to this asymmetrical following relationship, which encourages a more open discourse between users, along with its magnitude of users, choosing Twitter as the social media to analyse becomes the natural choice. Furthermore, unlike Twitter, other prominent social media platforms including Facebook and Instagram do not allow researchers open access to their data. We developed `MEP` to be able to design experiments, collect and analyse data from different Twitter APIs at scale in public cloud infrastructure.

Application Programming Interface. To work with and be able to analyse Twitter data efficiently on an arbitrarily large scale, access to Twitter's Application Programming Interface (API) is needed, and requires Twitter developer credentials, which anyone can apply for. With access to the credentials, one may request and download tweets which can be represented as `JSON`-files. At the time of writing, two versions of the Twitter API exists. This work was done in the older version 1.

[2] https://github.com/lamastex/mep.

[3] https://github.com/lamastex/spark-gdelt.

To get a sense of how the data was handled, a brief overview of the relevant fields from the schema of the `JSON` for a tweet will be presented. For full details, we refer to Twitter's data dictionary[4], [5]. The two most basic objects for a tweet are the User object and the Tweet object shown in Tables 1 and 2, respectively.

Table 1. Some attributes, with their types and description, for the User object (Table 1 of [24]).

User object		
Attribute	Type	Description
id	Int64	The unique integer representation of the user
screen _name	String	The screen name, also known as handle of the user
followers _count	Int	The number of followers the user has
friends _count	Int	The number of users the user follows

From the User object, as the name suggests, we get access to the metadata of a user. However, note that no direct information about which users follow the user, or which users the user follows, beyond the counts, is accessible from the user object.

From the Tweet object, we get access to the metadata of a tweet. Via the field "user", we also get the information of the user behind the tweet, since this is a User object.

Table 2. Some attributes, with their types and description, for the Tweet object (Table 2 of [24]).

Tweet object		
Attribute	Type	Description
created_at	String	UTC-time when the tweet was created
id	Int64	The unique integer representation of the tweet
text	String	The textual content of the tweet
in_reply_to_status_id	Int64	If the tweet is a reply to another tweet, the field will contain the tweet-ID of that tweet. Otherwise null
in_reply_to_user_id	Int64	If the tweet is a reply to another tweet, the field will contain the user-ID of that tweet. Otherwise null
user	User Object	All information of the user of the tweet
quoted_status	Tweet Object	If the tweet is a quote tweet, all information of the original tweet will be contained in this field. Otherwise null
retweeted_status	Tweet Object	If the tweet is a retweet, all information of the original tweet will be contained in this field. Otherwise null

[4] https://developer.twitter.com/en/docs/twitter-api/v1/data-dictionary/object-model/tweet.

[5] https://developer.twitter.com/en/docs/twitter-api/v1/data-dictionary/object-model/user.

Moreover, since the fields "quoted_status" and "retweeted_status" are Tweet objects, we get the full information of the original post that has been retweeted or quoted.

Note that the Tweet Object in "retweeted_status" points to the original tweet that has been retweeted, if the post is a retweet. It is possible for a user to retweet another user's retweet, but information on this chain of events is thus not accessible. For example, let user A write a tweet T that gets retweeted by user B. Later, user C sees this retweet on user B's timeline and then retweets T. Twitter's API will then only tell us that user B and C have retweeted user A, but not the fact that user C accessed this tweet via user B. This limitation also motivates the use of retweet network in Sect. 4.2.

Along with these two objects, there is another object named entities, which contains all the metadata of a tweet's content, including any URLs, hashtags, twitter handles of users mentioned, and media content (pictures and short video clips) (Fig. 1).

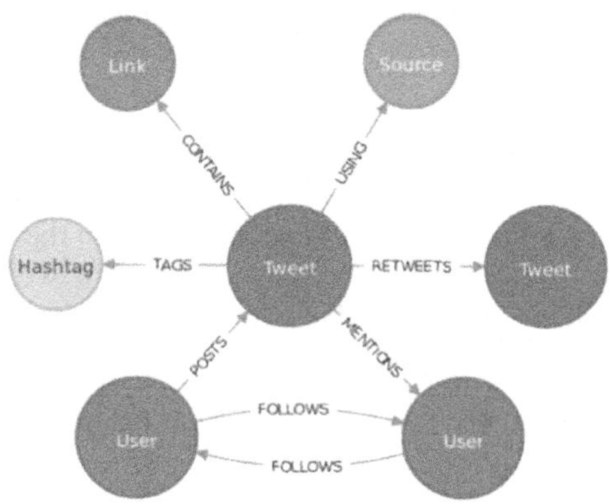

Fig. 1. An overview of Twitter's API. *Source* is simply which device (e.g., smartphone, and desktop) a user used to post the tweet. Note that included media in tweets are represented as links in this overview.

Data Set. The data set that was used [15] has 41.8 million collected tweets from 10.1 million unique users regarding the Black Lives Matter movement, along with the smaller counter movements of Blue Lives Matter (pro-police movement) and All Lives Matter. These tweets were collected by filtering on the keywords: *BlackLivesMatter*, *BlueLivesMatter* and *AllLivesMatter*. The data contains tweets from the beginning of the movement in 2013 to 30 June 2020. In this work, we focus on the events occurring during the aftermath of the death of George Floyd on 25 May 2020, and discard all tweets before this date.

Collecting Data. Due to Twitter's policy, collecting and sharing tweets publicly is not allowed. To share a set of tweets, instead one shares the IDs of each tweet, and to get the full metadata of the tweets, access to Twitter's API is needed. There is also a limit on how many tweets one may collect per hour, which initially was a problem. To get around this, the python library twarc[6] was used. twarc allowed us to collect tweets from the IDs (a process known as *hydrating*), in an optimised way with respect to the hourly collection limit.

To be able to work with the data in Databricks and Spark, a Docker-container with python and twarc was set up on a remote machine, that ran the hydration script on small batches of the IDs, collected them as '.json'-files, and then compressed and stored them in our Databricks cloud storage. This procedure took roughly five days.

A consequence of retroactively collecting tweets from their IDs is that all tweets that have been removed due to various reasons (such as the users of these tweets getting banned, removing their accounts, or going private) at the time of hydrating, are not accessible and were therefore not collected.

After hydrating the IDs from the data set, and discarding tweets posted earlier than 24 May 2020, 23.3 million tweets from 7.1 million unique users were left. These were cleaned to be easier to work with using Spark's Dataframes. We also categorised each tweet as an original tweet, retweet, quoted tweet, etc., and then stored them in the column-based data-storage format parquet on a delta lake [3]. See MEP for details of the collector, pre-processor and categoriser behind the delta lake.

3.3 GDELT

The *Global Database of Events, Language, and Tone* (GDELT) project, founded in 2013, is an open database supported by Google Jigsaw, that monitors news media in print, broadcast, and web formats from all over the world in over 100 languages. It is updated every fifteen minutes and stretches back to the 1st January of 1979, containing meta-data such as the people and organisations being mentioned, events and their locations, counts of key-words along with the tone and emotions of the parsed news sources[7]. We used the GDELT database to get a high level understanding of the mass media landscape during the given time span, by reducing the records of reported events of protests, to data points in time. We accomplish this by building an analytics-ready Delta Lake [3]. A brief overview of GDELT to appreciate how we handled the data for this work follows. For a more thorough overview, we refer to the documentation[8] and SPARK-GDELT , our open-source library developed for this study.

Coding. The idea behind GDELT is that of *coding*, which is fundamentally fairly simple. Given a *record* – for example a written news article – go through the text and identify the real world *events* that are being reported in the record, and identify the *actors* who are involved in the event. During the Cold war, two coding frameworks dominated: WEIS and the *Conflict and Peace Data Bank*, COPDAB. Both of these frameworks, being developed and used in a 20th century post-World War II context, were focused on codifying how sovereign states (the actors) interacted through official diplomacy and military threats [37]. For example, in the following sentence:

[6] https://twarc-project.readthedocs.io/en/latest/.
[7] https://www.gdeltproject.org/.
[8] http://data.gdeltproject.org/documentation/GDELT-Global_Knowledge_Graph_Codebook-V2.1.pdf.

*"**President Reagan** has **threatened** further action against the **Soviet Union** in an international television program beamed by satellite to more than 50 countries",*
one would identify the act of threatening as the event, and assign it some integer (decided by the code framework), with the actors being President Reagan (or the United States if the coder is only interested in sovereign states), and the Soviet Union.

This process of coding would historically be done by hand. However, the combination of psychological studies showing that the kind of sustained decision-making involved in coding leads to fatigue, inattention, and heuristic shortcuts, and the technological advancement in computing software and hardware, coding is nowadays automated. The frameworks for codifying has also developed since the cold war, with GDELT using the framework of *Conflict and Mediation Event Observations* (CAMEO) [23]. Some notable changes being that actors are no longer limited to sovereign states, and include persons, organisations, and companies.

In practice, GDELT is essentially two separate but interlinked databases: The *Global Knowledge Graph* (GKG), which consists of records and the *Event Database*, which as the name suggests stores events that are being reported.

GKG. The Global Knowledge Graph (GKG) consists of all records from multiple news sources in the world. As of version 2 of GDELT, new records get added every fifteen minutes. Whenever a record is added, the source text is parsed via natural language processing to identify the events (using coding), locations, persons and organisations, as well as themes mentioned in the text. Moreover, keywords such as "protest" that are mentioned multiple times gets counted. Sentiment analysis is also incorporated to get a value of the tone of the source text (whether the text is positive, neutral or negative). Many other metadata extracts are in each GKG record.

Event Database. The Event database attempts to record all unique events that are being identified in the parsing process of the GKG database. Each data point is given a unique ID for the event, and contains the date, the actors along with the code of the type of event being identified. The coded event also gets mapped to the Goldstein-scale [16], which seeks to measure the potential impact the event could have on the stability of the country. Moreover, the Event database has metadata on how often the event has been mentioned by records in GKG and the average tone of these records.

Handling of the GDELT Data. Due to the sheer magnitude of data contained in the GDELT database, working with data proved quite a challenge. Our goal was to filter out the events about the protests relating to the Black Lives Matter movement and the counter movements between 25 May 2020 and 30 June 2020. Although the parsing of news records into the GKG database identifies organisations, it did not identify the Black Lives Matter movement as one, probably due to its lack of centralisation.

What we did instead was to filter out all data relating to protests happening in the world. This naturally led to noisy data, since we got reports of protest unrelated to the BLM movement, but we justify this by the fact that no other major protests were happening in the world at the same time. To check this, we filtered the Event database

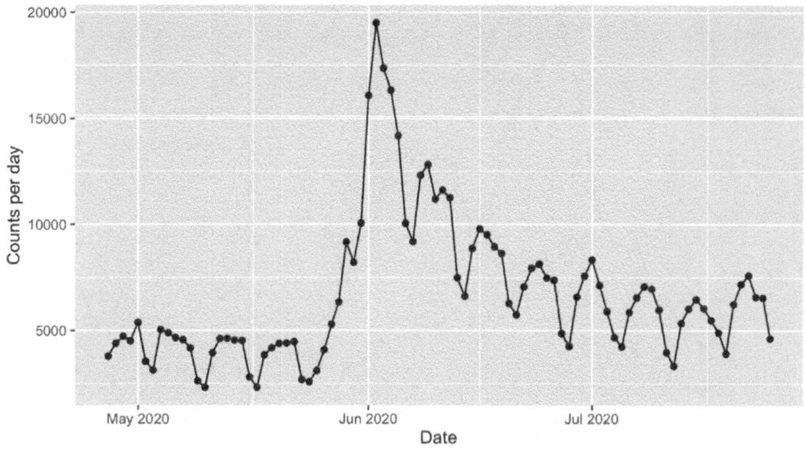

Fig. 2. Events coded as protests in the GDELT Event database (Fig. 1 of [24]).

by events with CAMEO root-code 14, i.e., those events coded as protests, over a three months timeline.

As we see in Fig. 2, there is a baseline of roughly 5,000 events per day coded as protests before 25 May. This number then explodes, and there is nothing that suggests that the sudden increase in magnitude of protests are not related to the BLM protests. It is worth pointing out that there is no bijection between the real world protest and the protest data from the Event database. For example, if in one city during one day, large protests are taking place and one group of people are protesting peacefully while another group is rioting, then the coding framework should identify the act of the peaceful and rioting protesters as two different events [37], although they are near each other in time and space. Thus, saying that more than 8,000 protests happened on the 1 June 2020, would be incorrect.

In Sect. 5 we will look at news reports in mass media, and therefore use data from the GKG database. We did this by filtering by the themes of the records. All records in the GKG database with theme "PROTEST" were filtered out.

Ignoring the periodic dips in the GKG plot in Fig. 3 (which are due to less reporting being done on weekends), the two plots follow a similar pattern. Naturally, there are more records than events, since multiple news sources may report the same event.

4 Analysis of Twitter Data

In this Section, we explore the Twitter data, first via simple querying on the data set, and then by doing network analysis on the induced retweet network. The results from this exploratory data analysis then motivated the choice of using Hawkes processes to model and perform hypothesis tests to shed light on the phenomena of interest in this study – occurrence of tweets in support of the BLM movement and that of mass media reports of street protests.

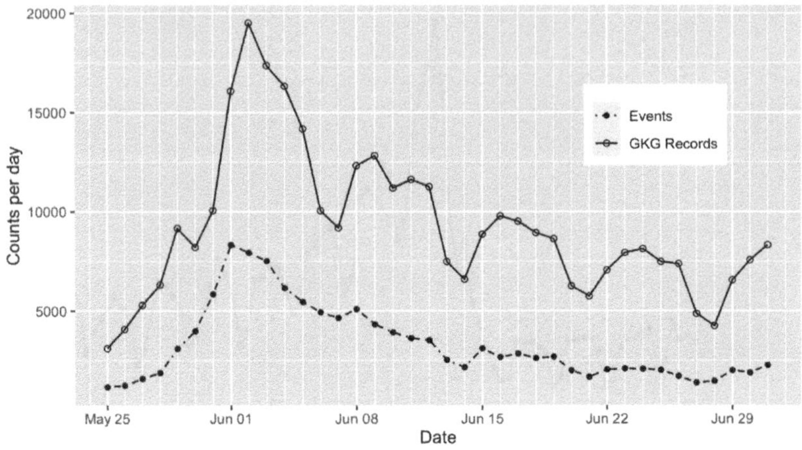

Fig. 3. Comparison of records from the GKG database with theme "PROTEST", and events coded as protests from the Event database (Fig. 2 of [24]).

4.1 Data Observations

Timeline. We started by examining the data over the relevant time-span from 24 May 2020 to 30 June 2020. During this period, 23,346,745 tweets by 7,111,140 unique users were collected using `twarc` on the BLM data set [15].

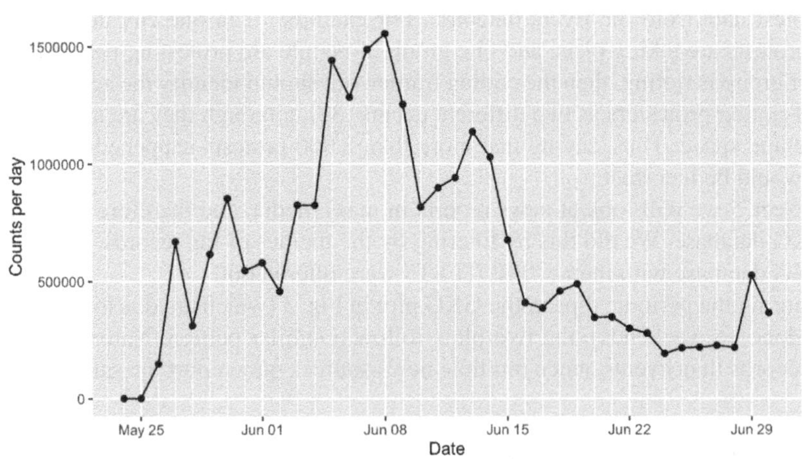

Fig. 4. Number of tweets per day (Fig. 3 of [24]).

From Figs. 4 and 5, we can see that activity first starts on Twitter, and the reports of protests start to drastically increase on 27 May. We also see a dip in Twitter activity between 31 May and 2 June, while the GDELT data on the number of reports of protests spikes during these days. The explanation of this is simply that the data set lacks tweets on these days. This was found while exploring the data, and noticing that the data set contained retweets of a tweet from this time period, but not the original tweet. Whether

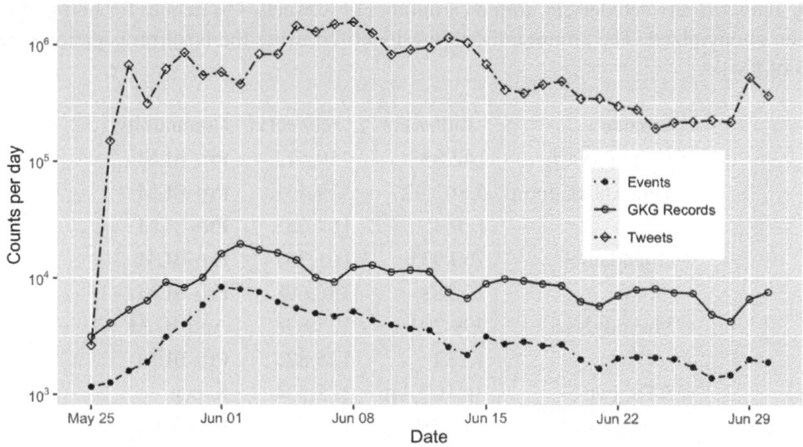

Fig. 5. Log-scaled plot of the number of tweets, records and events (Fig. 4 of [24])

these missing tweets disappeared during the collecting of data, or if they are missing in the original data set [15] of the Tweet IDs, remains unclear. To deal with this, we refrained from doing any modelling with tweets from this time period.

Type and Media Content of Tweets. Next, we examined TweetTypes, i.e., the types of status update or interactions in our Twitter data. The most to least frequent TweetTypes (% of data) were Retweets (55%), Retweets of Quoted Tweets (27%), Original Tweets (7%), Quoted Tweets (7%), Reply Tweets (3%), Original Tweets (1%). Thus, only 18% of the tweets in the BLM-data set were original tweets (either original, or replies to other tweets), with the remaining 82% being some sort of retweeted content. This suggests that the re-sharing of other users' original content is fundamental for how users interact with each other on Twitter, and motivated our choice of examining the retweet network.

One initial idea was to focus on URLs to news articles shared by Twitter users, and then link them to the GDELT database. However, we soon discovered that users in general did not share news sources from mass media. Instead highly retweeted tweets often contained original media (i.e., videos and pictures), which were often taken from the protests. For instance, 53% of tweets with over 1000 retweets, as opposed to only 17% of all tweets, shared original media.

4.2 Network Analysis

Section 4.1 showed the importance of retweets in the Twitterverse. In this Section we will formalise this by introducing a retweet network structure on our data set.

Retweet Network

Definition 10. *Let* $G_I = (V, E)$ *be a directed weighted graph in time interval* $I \subset \mathbb{R}_+$, *where every vertex* $v \in V$ *is a unique Twitter user, and every edge* $e \subset \{(u, v) \mid (u, v) \in E \subset V^2\}$ *is interpreted as user* v *having retweeted* u *during time interval* I. *The weight*

Table 3. Ten most retweeted users, sorted by number of retweets. Usernames for non-public users have been anonymized. The communities were identifed using the label propagation algorithm (Table 3 of [24]).

Username	| followers |	| retweets |	Community
@JoshuaPotash	142,833	759,572	Pro-BLM
@YourAnonCentral	5,862,927	529,431	Pro-BLM
–	1,584	187,065	Pro-BLM
@elijahdaniel	760,935	161,337	Pro-BLM
–	22,983	135,698	Pro-BLM
@MrAndyNgo	799,291	125,898	Anti-BLM
–	1,232	125,826	Pro-BLM
@BTS_twt	34,107,446	125,534	K-pop
@shawnwasabi	140,788	106,731	Pro-BLM
@Drebae_	141,613	103,594	Pro-BLM

$W(e) = W((u,v)) \in \mathbb{N}$ *is the number of times user* v *has retweeted user* u*. We then define* G_I *as a retweet network.*

Furthermore, we define G'_I *as an undirected retweet network if* $(u,v) \in E \Leftrightarrow (v,u) \in E$*. Thus* G'_I *ignores whether* u *retweeted* v *or vice versa but preserves the information that there is a retweet relation between the two users.*

We chose to look at retweets since a retweet by user u of an original tweet by user v is highly likely to mean that user u agrees with user v. Direct retweets are generally recognized to indicate trust in the communicator and endorsement [7,22,27]. The number of times a user has been retweeted also gives a probabilistic interpretation, using the random geometric graph interpretation in [35], that measures how influential a user is on another in terms of the lengths of their most retweeted paths.

By looking at our retweet network we can already get some information from the Twitter data set; simply by summing the outgoing edges and their weights for every user, we get the most retweeted users in our time interval between 24 May 2020 and 31 June 2020.

By just looking at the ten most retweeted users, there are a few points of interest worth mentioning. The first being that the third and the seventh most retweeted users both have fewer than 1600 followers. This naturally raises the question of exactly how these users got the exposure they did. By looking at the diffusion process of these users' tweets in Sect. 6, will we be able to draw any conclusions on how information can be spread in the Black Lives Matter movement?

One noteworthy user is the sixth most retweeted user @MrAndyNgo. Andy Ngo is an American conservative journalist and a prominent opponent of the Black Lives Matter movement, who in February 2021 published *Unmasked: Inside Antifa's Radical Plan to Destroy Democracy* [30], where he among other things writes about his experiences from the BLM protests of 2020. His presence amongst the most retweeted users will serve as a gateway into the counter-movements of All Lives Matter and Blue Lives Matter. Thus, we need to detect different communities within the observed retweet network, such that each community has more edges or retweets within it when compared to the number of edges between it and another community.

Connected Components. The motivation behind the definition of an undirected retweet network follows in the next step, when we look at the connected components of our graph.

Definition 11. *Let G be a graph. A sequence of edges* $(e_1, ..., e_{n-1})$ *is called path if it corresponds to a sequence of distinct vertices* $(v_1, .., v_n)$*, such that* $e_i = (v_i, v_{i+1})$. *Two vertices* u, v *are connected if there exists a path between them, and if G is undirected, we call the sub-graph H of G a connected component if and only if there exists a path between every pair of vertices in H which contains a subset of the vertices in G.*

The reasoning behind invoking the notion of connected components of the undirected retweet network is to, on a high level, make sure that a meaningful discourse between users, in terms of being influenced by and influencing others, exists within the connected component. In practice, we could have a very disconnected network with lots of unconnected components, which would mean that most users only interact and retweet a few selected users. Another interesting case would be if the network would have a few significantly large components; this would suggest the existence of a set of discourses, where the users in their respective component do not interact – perhaps because of political differences reflected in large "echo chambers". To find all connected components in the retweet network, the `GraphFrames` framework in Spark was used. The result showed that 6,083,687 i.e., 85.6% of the 7,111,140 users were in the same connected component. The remaining users were scattered around in smaller connected components, with the largest being 74 users. These users were therefore discarded from further analysis.

Community Detection. While the data set contains tweets using the hashtags of the counter movements #AllLivesMatter and #BlueLivesMatter, in practice, users associated with these movement did not necessarily use these hashtags, but often used the hashtag #BlackLivesMatter either ironically or to get more attention. Thus, just using simple querying on the hashtags in the data set, did not suffice to get a sample of users from those movements. To get a better sense of the relationship between users, we instead therefore used the community detection algorithm known as *Label propagation algorithm (LPA)*. LPA is a semi-supervised machine learning algorithm, which seeks to assign labels to nodes in a network, where each label maps to a specific community inside the network [33]. In Spark's GraphX framework, the algorithm is implemented using Pregel API [26], which allows for parallel computation when processing graphs. On a high level, Pregel computations are a sequence of iterations, defined as *supersteps*, where for every superstep, each vertex in the graph runs a user defined function. This local vertex-centric approach where each vertex is processed independently in parallel, in contrast to the more classical iterative graph algorithms where each vertex is visited one by one, naturally induces distributed implementations that can computationally scale to arbitrarily large networks. In distributed LPA, implemented as a Pregel program, each vertex in the graph is initially assigned its own distinct vertex label to represent its initial community label. At every superstep, vertices send their community label to all out-neighbours and update their label to be the mode community label of incoming messages from their in-neighbours. Although the algorithm can have trivial or oscillating solutions without guarantees on convergence, it works well in practice on real data as we found by running LPA on the largest connected component with 10 supersteps and investigating at least the most influential set of users within each community manually.

Exploring Ideological Diversity. By looking at the twenty most retweeted users, we see that eighteen of these fall into the same pro-BLM community, with 155,229 users. Andy Ngo is in a community with 26,624 users. This is interesting when we remind ourselves from Table 3 that he is the sixth most retweeted user, and if we assume that most of his retweets come from his relatively small community, it suggests that he has a very loyal set of core followers. The questions that arises then are if we can identify this core set of followers, and moreover if we also can identify a similar core followings in the pro-BLM community. In the same community where we find Andy Ngo, we also have prominent conservative commentators such as Candance Owens, Glenn Beck, Steven Crowder, Paul Joseph Watson, Dave Rubin, and also Republican senator Ted Cruz, and Raheem Kassam from the Reform UK-party (formerly known as The Brexit-Party), along with others. It is worth mentioning that all of the twenty most retweeted users in this community are users with largest followings (over 25,000 followers). Thus, the phenomena of users with small followings reaching a larger audience does not exist to the same extent in this community when compared to the pro-BLM community.

The last of the twenty most retweeted users is the official account of the South Korean pop (K-pop) group BTS, who has their own community. The communities for the top ten most retweeted users are presented in Table 3 and a sample of tweets from the pro-BLM and anti-BLM communities are presented in Table 4.

Note how the textual content of the tweets from the two communities differ. By going through the label propagation algorithm we seem to have identified the two different political camps. Moreover, we note that usage of the hashtag #BlackLivesMatter is prominent in the anti-BLM community. Thus, we can conclude that just filtering by the anti-BLM #AllLivesMatter and #BlueLivesMatter would not have sufficed to identify these communities.

Thus, through the use of (1) retweet network, which encodes retweets, one of the clearest signals of directional ideological concurrence of the retweeter with the tweeter, (2) distributed label propagation on such a retweet network to detect communities of users who are in ideological concurrence within each community, and finally (3) listing the top K most retweeted tweets within each such community, we have a simple yet effective mechanism to explore the *ideological diversity* that is representative of the communities, independent of their sizes and activity levels, i.e., the number of users and intensity of interactions in Twitter. We found this simple three-step process to be an effective approach to identifying the pro/anti-BLM tweets before further analysis.

4.3 Summary

In this chapter we have examined how the users in the data set have interacted by defining and looking at the retweet network. We noted that some of the users, despite their relatively small following, managed to become the most retweeted users. This motivates the questions:

Table 4. Sample tweets from the pro-BLM and anti-BLM communities (Table 4 of [24]).

Pro-BLM community
i can't stand by and continue to live in a world where the color of your skin is an automatic target on my family, friends, and neighbors backs. tri-city we must come together to support our communities. THIS. IS. AMERICA. BE THE CHANGE YOU WANT TO SEE. #blacklivesmatter https://t.co/XIDSNqgx6Q
Thread of people who took it upon themselves to trivialise the current situation going on and #BlackLivesMatter
#BlackLivesMatter Houston is hosting a protest march this FRIDAY at 2PM starting at Discovery Green demanding justice for #GeorgeFloyd White allies, y'all gotta do better and this is a place to start. Everyone who's able should be there. https://t.co /EbWeBrZneP
Aiyana Jones a 7 YEAR OLD CHILD who was shot in the head by an officer, when the officer raided the wrong house. A 7 year old girl didn't deserve to be killed because of disgusting reckless officers. Acab and BLM, never forget this girls name! #BlackLivesMatter https://t.co/HCWzabkFv4
So protest in Huntsville, TX was small, but that was no surprise. We're a small town and most things just caught up to the present on the outside...at the end of the protest on my way home, I saw something I never noticed. This is why we do what we do. #BlackLivesMatter https://t.co/gTuCilB7mi
Anti-BLM community
Black people are 80 times more likely to kill white people in England/Wales than the reverse! And yet, #BlackLivesMatter more than others? EXPLAIN... Check the stats: https://t.co/DmPDVVGbSo https://t.co/qxXmuNIh2X
#BlackLivesMatter should now be classified as an extreme political hate group.. Simple.. https://t.co/mFh56qCpo9
#DontTakeTheKnee #DontTakeTheKnee please get this trending Sick & tired of the #ScumMedia telling us what we should do! Well I say #DontTakeTheKnee #BLM is a terrorist organisation. Do your homework! #AllLivesMatter #WhiteLivesMatter #ISTANDwithDominic Raab @SkyNews
Then someone gets stabbed and they want the police back after running them out of town. Ha you couldn't make it up #BlackLivesMatter #blm #thugs #brixton https://t.co/1uVXQ63UT2
Just saw a video of #BlackLivesMatter protest in #Reading - looks like 3 white people have been stabbed and in a bad way! Now if this turns out to be a race attack, I'm going to blame the #Media. They've been stoking up tensions between blacks and whites for weeks now!

- What is the nature of the information diffusion process for retweets in the data set in general? Does tweets from different movements spread in the same way?
- In particular, how does the diffusion process look for a viral tweet when it is initialized by a user with a small following?

We also implemented a community detection algorithm to find groups of users who share similar values. This proved quite successful, by looking at the users, and sampling the textual content of the tweets from these communities. The fact that the sixth most retweeted user Andy Ngo belonged to a relatively small community, raises the question:

- What role does a influential users play in the spread of a tweet? Can such influential users be found for all viral tweets?

These are the main questions that motivates the choice of models in Sect. 6.

5 Joint Media Modeling

In this Section we examined the interplay between the Twitter and GDELT data sets by looking at the Granger causality between them. For this we proposed simple two-dimensional Hawkes processes with an exponential kernel. The timeline for this joint modeling was three days after the death of George Floyd over the 24-hours-long period between midnight of 28 May and midnight of 29 May, which is when the protests had just started to spread nationwide across the US, and also become violent.

5.1 Model and Data

In dimension one we had the Twitter data. To control the magnitude of the data we only considered original tweets, i.e. all retweets were filtered out, that had at least one retweet, to filter out tweets made by users with a negligible following. Moreover, we examined the 20 largest communities and identified one anti-BLM (the same community identified in the previous section), and filtered out all tweets made by users from that community, so that we only considered pro-BLM tweets. This left us with $10,774$ tweets.

In the second dimension we had records from the GKG-database from GDELT. The records were first filtered on mentioned themes, and only those reporting events of protests were selected. This naturally lead to some noise in the data, due to not being able to precisely filter out only the events mentioning protests relating to the Black Lives Matter-movement. To reduce this noise, we also filtered on records that mentioned George Floyd. While in theory a record could report a BLM related protest without mentioning George Floyd, we reasoned that since our timeline of interest was three days after his passing, most records should mention George Floyd to give the reader some context for the reported protest. To handle that the GKG-database updates in intervals every 15 minutes, every record got a randomised timestamp in the fifteen minute interval prior to it being added into the database, to get the records in continuous time. With this query in the selected time interval, $3,341$ records were found.

Given this data, we jointly model events in social and mass media by fitting the multivariate Hawkes process in Definition 7. We want to test whether or not Granger causation exists between dimensions 1 and 2 representing events in Twitter and events in mass media from the GDELT project, respectively. As per Theorem 2, parameter $\alpha_{12} = 0$ if and only if mass media events do not Granger cause Twitter events, and vice versa for $\alpha_{21} = 0$.

5.2 Results

The data was fitted using `python` library `tick`[9]. `tick` requires that the decay parameters β_{ij} are given as constants beforehand, which then allows highly efficient fitting of the remaining parameters μ_i and α_{ij}, using accelerated gradient descent [6]. The problem of fitting the decay parameter β in the exponential kernel is well-known [36], and is due to the fact that while the baseline parameter μ and excitation parameter α

[9] https://x-datainitiative.github.io/tick/.

can be efficiently computed via convex optimisation, this is not always true for β. With this in mind, we proposed three different models where the decay parameters β_{ij} were handled differently:

- $\mathcal{M}_0$: $\beta_{ij} = 1,\ \forall(i,j) \in \{1,2\} \times \{1,2\} =: \{1,2\}^2$
- $\mathcal{M}_1$: $\beta_{ij} = \beta \in (0,\infty),\ \forall(i,j) \in \{1,2\}^2$
- $\mathcal{M}_2$: $\beta_{ij} \in (0,\infty),\ \forall(i,j) \in \{1,2\}^2$

To compare the different models, we looked at (i) the *Akaike information criterion* AIC $= 2k - 2\ln(\hat{L})$, where k is the number of estimated parameters, and $\hat{L}$ is the maximum likelihood of the model, (ii) the *relative likelihood* $\exp((\text{AIC}_p - \text{AIC}_q)/2)$, where the AIC values for models p and q satisfy $\text{AIC}_p < \text{AIC}_q$, and (iii) the likelihood-ratio test statistic $\lambda_{\text{LR}} = -2\ln(\hat{L}_p/\hat{L}_q)$.

Comparison Between $\mathcal{M}_0$ and $\mathcal{M}_1$. Setting $\beta_{ij} = 1$ for all i, j in model $\mathcal{M}_0$ gave us the log-likelihood value of 372.981, and AIC $= -733.963$ (where $k = 6$ for the two estimated baseline parameters μ_i and the four excitation parameters α_{ij}. For model $\mathcal{M}_1$, we did a sequential grid-search over β's, by using the convex optimiser in `ticks` to quickly obtain the most likely μ_i and $\alpha_{i,j}$'s for each fixed $\beta_{i,j} = \beta$, to find the most likely parameter $\hat{\beta} = 6.17$, with the maximum log-likelihood value of 384.771 and AIC $= -755.542$ (where $k = 7$ since we now also estimate β). Note that the known problem [36] of β misbehaving and being noisy and fluctuating with respect to the likelihood of the model, did not seem to occur with our data possibly due to (i) the structure or information of our data, as well as, (ii) our simplectic parametrisation of $\mathcal{M}_2$: $\sum_{ij} \beta_{ij} = 4\hat{\beta}$, where $\hat{\beta}$ is the value from model $\mathcal{M}_1$ with highest likelihood, for numerically stable optimisation.

The relative likelihood of the models was 2.0624×10^{-5}, i.e., model $\mathcal{M}_0$ was 2.0624×10^{-5} times as probable as model $\mathcal{M}_1$ to minimize the information loss. Since $\mathcal{M}_0$ is nested in $\mathcal{M}_1$, i.e., the parameter space of $\mathcal{M}_0$ is a proper subset of that of $\mathcal{M}_1$, we do a likelihood ratio test and reject $\mathcal{M}_0$ in favour of $\mathcal{M}_1$ ($\lambda_{\text{LR}} = 23.5781$, p-value $< 10^{-7}$).

Comparison Between $\mathcal{M}_1$ and $\mathcal{M}_2$. Model $\mathcal{M}_1$ and $\mathcal{M}_0$ assume that the decay parameters β_{ij}'s are identically $\beta \in (0,\infty)$, i.e., the decay parameter within each dimension and between every pair of dimensions is given by the same value. The real-world interpretation of this is that tweets and mass media reports stay relevant for the same amount of time into the future, which seems like a major assumption as mass media dissemination and social media communication are fundamentally different in nature. To account for this, we introduced model $\mathcal{M}_2$, where each β_{ij} can vary freely in $(0,\infty)$.

We did a sequential grid search over the 4-simplex, similar to the one-dimensional case of $\mathcal{M}_1$, to find the most likely β_{ij}'s, except that we used the constraint that all β_{ij} should lie in the interior of the 4-simplex, induced by $\hat{\beta}$, such that $\sum_{ij} \beta_{ij} = 4\hat{\beta}$, where $\hat{\beta} = 6.1700$. Iterating through 32509 values we found the most likely values to be $\hat{\beta}_{11} = \hat{\beta}_{22} = 16.170$, $\hat{\beta}_{12} = 3.702$, and $\hat{\beta}_{21} = 8.638$, at the maximum log-likelihood value of 384.772, with $k = 10$ and AIC $= -749.544$. Note that despite having three additional parameters, the maximum log-likelihood of $\mathcal{M}_2$ is close to that of $\mathcal{M}_1$, with the relative likelihood of the models, likelihood-ratio test statistic, and p-value being 0.04984, 0.002121, and 0.9971, respectively. We therefore do not reject $\mathcal{M}_1$ in favour of $\mathcal{M}_2$ and choose $\mathcal{M}_1$ for further analysis.

Fitting the Data Using $\mathcal{M}_1$. To find whether Granger causality between the two dimensions exists, we were interested in whether parameters $\hat{\alpha}_{12}, \hat{\alpha}_{21}$ are equal to 0 or not. Fitting the data using model $\mathcal{M}_1$ with estimated decay parameter $\hat{\beta} = 6.1700$ gave us the following estimated parameters $\hat{\mu}_1 = 1.000$, $\hat{\mu}_2 = 0.998$, $\hat{\alpha}_{11} = 0.986$, $\hat{\alpha}_{12} = 0.0327$, $\hat{\alpha}_{21} = 0.0216$, $\hat{\alpha}_{22} = 0.921$. Note that the point estimates satisfying: $\hat{\alpha}_{12} > \hat{\alpha}_{21} > 0$, implies that there exists Granger causality between reported protests and tweets regarding the BLM-movement, provided we account for the errors in their estimation, i.e., their confidence intervals. We address this next using non-parametric bootstraps.

Hypothesis Testing. The following null hypotheses were proposed:

- $H_{0,12} : \alpha_{12} = 0$, i.e., reports of protests in mass media do not Granger-cause communication events in Twitter related to the BLM-movement.
- $H_{0,21} : \alpha_{21} = 0$, i.e., communication events in Twitter related to the BLM-movement do not Granger-cause reports of protests in mass media.
- $H_0 : \alpha_{12} = \alpha_{21} = 0$.

To get the confidence intervals for α_{12}, α_{21} we did a non-parametric bootstrap by sampling the observed data with replacement, and then estimating the parameters on the bootstrapped data under model $\mathcal{M}_1$. This was repeated 1000 times.

For α_{12}, i.e., the influence of mass media on Twitter, the 99-th percentile bootstrapped confidence interval is $(0.000, 0.09405)$, and therefore we cannot reject the null hypothesis $H_{0,12}$ that $\alpha_{12} = 0$ by the Wald test. Thus, the reports of street protests in mass media do not Granger-cause the pro-BLM interactions in Twitter.

On the other hand, the 99-th percentile bootstrap confidence interval for the parameter α_{21} that models Twitter's influence on mass media is $(0.01479, 0.02949)$, and therefore we reject the null hypothesis $H_{0,21}$ that $\alpha_{21} = 0$ by the Wald test. Thus, the pro-BLM interactions in Twitter Granger-cause the reports of street protests in mass media. We therefore also reject the common null hypothesis that there is no Granger causality whatsoever between social and mass media events around the BLM-movement, i.e., $H_0 : \alpha_{12} = \alpha_{21} = 0$.

To estimate type I error, i.e., the probability of rejecting the null hypothesis H_0, when it is true, we simulated data from the null hypothesis H_0, i.e., from the most likely parameters in $\mathcal{M}_1$, while restricting $\alpha_{12} = \alpha_{21} = 0$. For each such simulated data, we then performed the Wald test using non-parametric bootstraps by sampling the data with replacement $1,000$ times. Only one out of 100 such simulations from H_0 was rejected giving 0.01 as the Monte Carlo estimate of the Type I error.

6 Modelling Retweet Cascades

In this Section we will model the diffusion process of a retweet cascade, given one initial tweet. For this we will use marked Hawkes processes with the power-law kernel introduced in Sect. 2. The motivation for using a marked Hawkes process stems from the following properties in a retweet cascade, that we seek to capture:

- **Word-to-Mouth Spread:** When a user shares a tweet, the tweet will organically find its way into a new set of users, and from them into another new set of users, and so on.
- **The Magnitude of Influence:** Users with more followers tend to get more retweets.
- **Memory Over Time:** Most of the retweeting by users happen when they first see it in their timeline.
- **Content Quality:** The better a tweet is, vaguely speaking, the more retweets it will get.

Let us now look at the intensity of a marked Hawkes process with a power-law kernel on $\mathbb{R}_+ \times \mathbb{N}$, where each point (t_i, m_i) is a retweet at time t_i and where the mark m_i is the number of followers of the user who retweets.

$$\lambda(t) = \sum_{(t_i,m_i)\in\mathcal{H}_{t,m}} \kappa m_i^{\beta}(t - t_i + c)^{-(1+\theta)}. \tag{8}$$

The property of word-to-mouth is naturally captured by the self-exciting nature of Hawkes processes; when a tweet is shared, a new set of users get access to this tweet in their time-line, and the intensity of the process, i.e., the probability of a new retweet, will increase. The magnitude of influence is captured by the fact that we are implementing a marked Hawkes Process. Since we let m_i equal the number of followers of the user who retweets, it follows that users with larger followings will contribute to a larger jump in intensity, scaled by parameter β. Since $(t - t_i + c)^{-(1+\theta)} \to 0$ as $t \to \infty$, the kernel is monotonically decreasing, and the property of memory over time is taken care of. Furthermore, κ scales the quality of a tweet, such that, larger values of κ result in larger jumps in intensity. The motivation for the requirement of this property is that we could have a relatively large retweet cascade, from a user without a significantly large following, that results in a large spread.

All of these properties are captured by other marked kernels, and we did some exploratory fitting using a marked exponential kernel. However, the marked power-law kernel gave best results as confirmed by related works [28,41].

6.1 Fitting Marked Hawkes Processes

The marked Hawkes processes with a power-law kernel has four parameters $\boldsymbol{\theta} = \{\kappa, \beta, c, \theta\}$. We estimate these by computing the maximum likelihood. The log-likelihood for the intensity is of the following form:

$$
\begin{aligned}
\mathcal{L}(\kappa, \beta, c, \theta \mid \mathcal{H}_{t_n}) &= \log\mathbb{P}(\{(m_i, t_i), i = 1, ..., n\}) \\
&= \sum_{i=1}^{n} \log\left(\lambda(t_i)\right) - \int_0^T \lambda(\tau)\mathrm{d}\tau \\
&= \sum_{i=2}^{n} \log\kappa + \sum_{i=2}^{n} \log\Bigg(\sum_{t_j<t_i} \frac{m_j^\beta}{(t_i - t_j + c)^{1+\theta}} \Bigg) \\
&- \sum_{i=1}^{n} \int_{t_i}^{T} \kappa m_i^\beta (t - t_i + c)^{-(1+\theta)} \mathrm{d}t \\
&= \sum_{i=2}^{n} \log\kappa + \sum_{i=2}^{n} \log\Bigg(\sum_{t_j<t_i} \frac{m_j^\beta}{(t_i - t_j + c)^{1+\theta}} \Bigg) \\
&- \kappa \sum_{i=1}^{n} m_i^\beta \left[\frac{1}{\theta c^\theta} - \frac{(T + c - t_i)^{-\theta}}{\theta} \right].
\end{aligned} \tag{9}
$$

The term $\int_0^T \lambda(\tau)\mathrm{d}\tau$ is a normalisation factor that we get by integrating the event rate over the time interval $(0, T)$. This is non-linear optimisation problem was solved numerically using the `R` library `evently`[10], which builds on `AMPL`[11]. Due to the requirement of `AMPL`, we implement `evently` in a remote machine via a Docker container with `R` and all required packages, in order to obtain the maximum likelihood fit of the data.

Results. The remote machine was set up with 64GB of RAM due to our compute budget constraints. However, when trying to fit retweet cascades with more than roughly 3000 retweets, the machine ran into lack of memory errors. Note that the largest retweets cascades in ourdata set had around 300000 retweets. These were certainly rare, but would have been interesting to study in more detail. We should however point out that cascades of around 3000 are still relatively big, and certainly big enough to model diffusion processes of interest.

Fitting was done on 20 retweet cascades initialised by users from both the BLM and anti-BLM communities identified in Sect. 4.2. By just comparing the plots of the intensities over time, and the fitted parameters, no clear distinction between cascades from the two communities could be made.

However, in both communities, retweet cascades of the type (Fig. 7) where the intensity suddenly spikes when an influential user joins the diffusion process, were found.

This phenomena is the most probable explanation for how some of the largest cascades in thedata set that were initialised by users with relatively small followings managed to diffuse to a large group of users (Fig. 6).

[10] https://github.com/behavioral-ds/evently.

[11] https://ampl.com/.

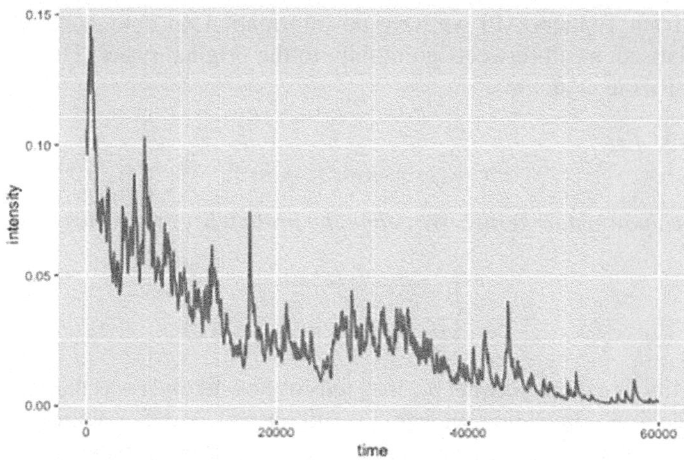

Fig. 6. Cascade initialised by a relatively highly influential user.

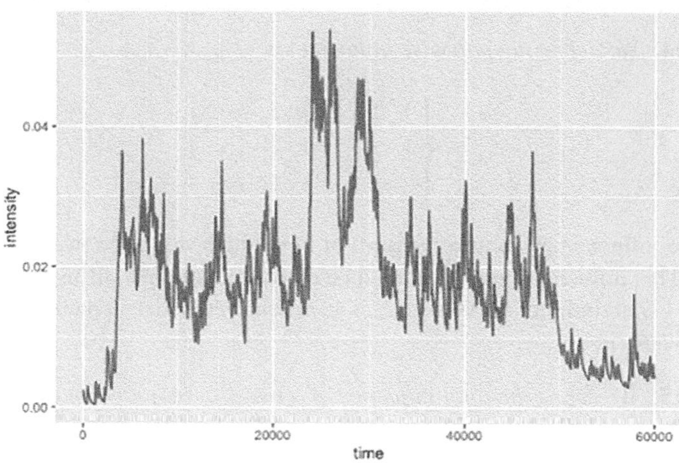

Fig. 7. Cascade initialised by a non-influential user.

6.2 User Influence

Taking the results from fitting different types of cascades, and discovering the type of information diffusion where a user with a large following joins the cascade at a later time, we here present an idea on how to recreate a probable branching process of the retweet diffusion, using the fitted parameters of the kernel in our marked Hawkes process.

Definition 12. *Given a retweet cascade, let a diffusion scenario G be a directed tree, where for each* $v_i, v_j \in G$, v_i *has an edge to* v_j *if the retweet* v_j *is a direct retweet of* v_i.

Recall that from Twitters API we have no information on how the actual branching process initialised, as all retweets point only to the original tweet. The following idea comes from Rizoiu et al. [34].

Definition 13. *Let*

$$\phi^p(m_i, t) = \kappa m_i^\beta (t + c)^{-(1+\theta)}, \tag{10}$$

be the marked power-law kernel. We define the probability of direct retweet as:

$$p_{ij} = \begin{cases} \frac{\phi^p(m_i, t_j - t_i)}{\sum_{k=1}^{j-1} \phi^p(m_k, t_j - t_k)} & i < j, \\ 0 & i \geq j. \end{cases} \tag{11}$$

The probability of a direct retweet p_{ij} thus tells us how likely it was that the j:th retweet in the cascade was a direct retweet from the i:th retweet, and thus a direct descendent in the diffusion scenario G. Since we have the marked kernel from the Hawkes processes in the definition, naturally retweets close to each other in time will have a higher probability of being direct retweets. The definition also takes into consideration that users with more followers will have a larger probability of getting direct retweets.

Definition 14. *We define the pairwise influence as:*

$$r_{ij} = \begin{cases} \sum_{k=1}^{j-1} r_{ik} p_{kj} & i < j, \\ 1 & i = j, \\ 0 & i > j. \end{cases} \tag{12}$$

The pairwise influence gives us a measure of how much influence tweet v_i exerts over retweet v_j. This influence can either be via a direct retweet in a diffusion scenario, but also when v_j is an indirect retweet of v_i, i.e., when there exists a path from v_i to v_j in a diffusion scenario.

Definition 15. *We define the total influence of a tweet v_i in a retweet cascade as:*

$$\varphi(v_i) = \sum_{k=1}^{n} r_{ik}. \tag{13}$$

We now have a measure of how influential users are in a retweet cascade. Note that while we are talking about the influence of a tweet, we are actually taking the heuristic shortcut and assuming that this is the influence of the user of that tweet; the reasoning behind defining the influence in terms of tweet influence, is that technically a user may retweet the same tweet more than once, and these retweets will then be given different total influence. In Table 5 we present an example from a retweet cascade.

Note that the user who joins the retweet cascade at time 1550.91 is given an influence score of 22.69, although this user has quite a large following of 51256 users. Compare this to the user who joins the cascade at time 565.05, with a following of 16527 users. Due to the fact that this user joins the cascade earlier, their influence score is much higher at 143.13, even though their following is nearly three times as small.

Table 5. The 8 most influential users in a retweet cascade.

Time	Number of followers	Total influence
0.00	1475	195.00
621.00	142881	161.28
565.05	16527	143.13
165.38	2285	118.81
304.53	591	27.39
737.97	27081	24.18
1550.91	51256	22.69
544.08	546	20.32

6.3 Summary

In this Section we modelled retweet cascades using the `R` library `evently`. Due to the limitations of hardware and lack of computational efficiency when fitting these models, we were only able to look at retweet cascades with magnitude of roughly 3000 retweets. From the fitting of retweet cascades, two distinct patterns in the diffusion process were identified; one where the intensity of Hawkes process starts high and peaks close to the initialisation of the retweet cascade, and one where the intensity reaches its peak at point later in time after its initialisation.

Influenced by these two patterns, we introduced a heuristic method of measuring influence scores of users in a retweet cascade, by recreating a probabilistic diffusion process retroactively via the fitted Hawkes process. While no major conclusions from this will be drawn in this Section, our hope is that these models can be used as tools in other fields, e.g., social sciences, for doing comparative research jointly with insights from field and/or online ethnography. In an ideal setting, with a better scalable estimator for the parameters at hand or with a larger compute budget for single machine with more memory, we would be able to fit all retweet cascades in our data set, which in turn would allow us to do more interesting analysis, such as looking at the distribution of the parameters and also identifying the most influential users for the whole data set, by summing and normalising the users' influences from all cascades.

6.4 Related Work

The idea of using Hawkes processes to model information diffusion and in particular retweet cascades is well established, often in the context of predicting, the final size of a retweet cascade given the first initial tweets in a given time interval. In this Subsection we present brief overviews of some related work.

A well known model for prediction is the SEISMIC (Self-Exciting Model of Information Cascades) model by Zhao et al. [41]. The SEISMIC model also implements a marked Hawkes process with a power law kernel, but unlike our approach, they fix parameters for the kernel for all retweet cascades. Instead they extend the Hawkes process by introducing a *infectiousness* parameter p_t, which aims to model how likely a post is to be re-shared at time t.

The intensity for their point process is of the form:

$$\lambda(t) = p_t \cdot \sum_{(t_i, m_i) \in \mathcal{H}_{t,m}} \phi_{m_i}(t - t_i), \tag{14}$$

which is a Cox-process (i.e., a double stochastic process), since the infectiousness p_t is stochastic. The main idea for SEISMIC is then to, given a retweet cascade at some fixed time t (e.g., $t = 60$ minutes) find an estimate of the infectiousness $\hat{p}_t$ (which is bounded by $\hat{p}_t < \frac{1}{\nu}$, to not explode exponentially and where ν is the branching factor of the Hawkes process) and then from $\hat{p}_t$ estimate the final size of the retweet cascade.

Mishra et al.'s [28] approach – which was our main source of inspiration for the modelling done in this Section – also presents models for prediction, and argues that the kernel of the Hawkes process should not be fixed, since retweet cascades may die out quickly or slowly and still end up with roughly the same number of retweets. Therefore, they fit each cascade in a similar fashion as was done by us using a non-linear solver. Moreover, they add a predictive layer using the estimated parameters as features and then train a Random Forest regressor, to predict the final size of the cascade after a given initial time. They also introduce a purely feature driven predictor with basic user features (number of followers, number of posts etc.), temporal features (waiting time for posts in the cascade), volume (i.e., the size of the cascade at the given time), past user success (average size of cascades previously initialised by the given user), and – perhaps more interestingly in the context of this work – a hybrid model, which combines the Hawkes process with a predictive layer, and the purely feature-driven predictor.

All three of these models outperform the SEISMIC model when tested on two data sets (where the first one is the same data set used in Zhao et al. for the SEISMIC model) containing cascades of size greater or equal to 50, with 30463 and 110954 cascades, respectively, and mean length of 160 and 158, and median of 95 and 90, respectively. Note that most cascades are therefore rather small compared to the cascades we managed to fit from the BLM data set. Moreover, all predictors based on Hawkes processes will fail when the conditional intensity at the time of prediction is too high, i.e., when the branching factor $\nu \geq 1$ in Mishra's models, and when the infectiousness parameter $p \geq \frac{1}{\nu}$. This suggests that there is an upper bound on the magnitude of retweet cascades, when successfully modelling via such Hawkes processes.

In Zhou et al. [42] a multivariate Hawkes process is implemented to discover the social influence of users in a network. In the article, a multivariate Hawkes process in 500 dimensions, each representing a popular website, is fitted on a data set containing timestamps of the event that one of the 500 sites creates a hyperlink to another site, to find the most influential sites that are quickest in detecting trends on the internet, and moreover the community structure of these sites.

The main idea is to let every user $u = 1, \ldots, U$ in the data set be represented by one dimension in a multivariate Hawkes process in U dimensions. The estimated baseline intensities μ_u, and excitation parameters α_{uv} (which captures the influence of excitation from dimension v to u) are collected into matrix form $\boldsymbol{\mu} = \mu_u, \boldsymbol{A} = \alpha_{uv}$, where $\boldsymbol{A}$ is defined as the infectivity matrix. To estimate the parameters in the Hawkes process, a low-rank and sparse regularisation is imposed on $\boldsymbol{A}$. The motivation for this is that the sparsity of the infectivity matrix $\boldsymbol{A}$ captures the fact that most users only influence a small number of users, while there can be a few very influential users. The low-rank structure of $\boldsymbol{A}$ is meant to capture the structure of communities in the network of users, where we interpret a set of linear dependent column vectors as a community. An algorithm for solving the optimisation problem of estimating $\boldsymbol{A}$ so that it is both low-rank and sparse is presented in their article.

This approach should in theory work for Twitter data and its restrictions given by the API, and could then be another way of identifying influential users and community structures. The data would then be tweets represented as a tuple of a timestamp and user, with no regard to whether they are original tweets or retweets. Other features such as how many followers a user has would also be discarded. The estimated influence of a user u in the data set, would then be given by the u:th column in the low-rank and sparse estimated infectivity matrix $\boldsymbol{A}$. We note that our approach to community detection in Sect. 4.2 is complementary to this approach as it uses the retweet network directly via distributed label propagation to identify ideologically diverse communities in a manner that allows to drill down to individual influential tweets within each community, while allowing for arbitrary sizes or the activity levels within each community.

7 Conclusion

We set out to ask the most basic question of whether peoples still have power in taking collection real-world action despite the concentrated ownership of mass media with their propaganda model of communication and the algorithmic attention rent seeking business models of corporate social media platforms, albeit to a lesser extent in Twitter during the summer of 2020 compared to its current revised business model. We find with some optimism that the answer is yes and peoples' power seems to be alive and kicking from our statistical hypothesis testing and analysis of social and mass media events around the Black Lives Matter (BLM) movement and its counter-movements during the summer of 2020 using mathematical models conducive to causal studies.

More specifically, we jointly model and test hypotheses about causal relationships between interactions in social media and the reports in mass media during the BLM protests following the death of George Floyd, by implementing open-source pipelines through `MEP` and `SPARK-GDELT` to process the data, i.e., extract, load, transform, explore, from scratch and at scale, on cloud infrastructure, and by employing self-exciting Hawkes processes and their Granger causal inference machinery.

We reject the null hypothesis that there is no causal relationship, and show that communication events in Twitter, surrounding tweets that supported the BLM movement, Granger-caused the reports of street protests in mass media from the GDELT project. However, we cannot show that the reporting of street protests in mass media Granger-caused the corresponding communication events in Twitter. We identified such pro-BLM tweets thorough a network analysis of the Twitter data to identify communities of users who have a shared ideology among an ideologically diverse set of communities. We also model social media diffusion process using specialised kernels towards understanding how and when information is spread influentially as cascades into the socially networked communities.

We thus establish a verifiable causal relationship between social media interactions in Twitter that are supportive of the global BLM social movement on one hand, and global mass media reports of street protests in solidarity with the movement on the other. This suggests that activists have harnessed social media to raise awareness and mobilise street protests in a one-way causal relationship whereby pro-BLM social media communications Granger-cause the reports of street protests by mass media.

Acknowledgements. AL was supported by a summer internship at Combient Competence Centre for Data Engineering Sciences. SL and RS were partially supported by Swedish Research Council project no. 2019-03351 and RS was partially supported by the Wallenberg AI, Autonomous Systems and Software Program funded by Knut and Alice Wallenberg Foundation. Computing infrastructure was supported by Databricks University Alliance and AWS.

References

1. Anad, B., Di Tella, R., King, G., Legg, H.: The future of media project: US media ownership. https://projects.iq.harvard.edu/futureofmedia/us-media-index. Accessed 10 Nov 2023
2. Aral, S.: The Hype Machine: How Social Media Disrupts Our Elections, Our Economy and Our Health – and How We Must Adapt. HarperCollins Publishers (2020). https://books.google.se/books?id=GlXODwAAQBAJ
3. Armbrust, M., et al.: Delta lake: high-performance acid table storage over cloud object stores. Proc. VLDB Endow. **13**(12), 3411–3424 (2020). https://doi.org/10.14778/3415478.3415560
4. Armbrust, M., et al.: Spark SQL: relational data processing in spark. In: Proceedings of the 2015 ACM SIGMOD International Conference on Management of Data, SIGMOD 2015, pp. 1383–1394. Association for Computing Machinery, New York (2015). https://doi.org/10.1145/2723372.2742797
5. Bacry, E., Mastromatteo, I., Muzy, J.F.: Hawkes processes in finance. arXiv (2015)
6. Bacry, E., Muzy, J.F.: First- and second-order statistics characterization of Hawkes processes and non-parametric estimation. IEEE Trans. Inf. Theory **62** (2016)
7. Boyd, D., Golder, S., Lotan, G.: Tweet, tweet, retweet: conversational aspects of retweeting on Twitter. In: Proceedings of the 2010 43rd Hawaii International Conference on System Sciences, HICSS 2010, Washington, DC, USA, pp. 1–10. IEEE Computer Society (2010). https://doi.org/10.1109/HICSS.2010.412
8. Buchanan, L., Bui, Q., Patel, J.K.: Black Lives Matter May Be the Largest Movement in U.S. History. New York Times (2020). https://www.nytimes.com/interactive/2020/07/03/us/george-floyd-protests-crowd-size.html
9. Daley, D., Vere-Jones, D.: An Introduction to the Theory of Point Processes: Volume I: Elementary Theory and Methods, 2nd edn. Springer, Cham (2003)
10. Deliso, M.: ABC News - Timeline: The impact of George Floyd's death in Minneapolis and beyond. ABC News (2021). https://abcnews.go.com/US/timeline-impact-george-floyds-death-minneapolis/story?id=70999322
11. Eichler, M., Dahlhaus, R., Dueck, J.: Graphical Modeling for Multivariate Hawkes Processes with Nonparametric Link Functions. Probability Theory and Related Fields (2012)
12. Falcon, A.: Aristotle on Causality. The Stanford Encyclopedia of Philosophy (Spring 2019 Edition) (2019)
13. Filimonov, V., Sornette, D.: Apparent criticality and calibration issues in the Hawkes self-excited point process model: application to high-frequency financial data. Quant. Financ. **15** (2016)
14. Frenkel, S., Kang, C.: An Ugly Truth: Inside Facebook's Battle for Domination. Little, Brown Book Group (2021). https://books.google.se/books?id=0uEMEAAAQBAJ

15. Giorgi, S., Guntuku, S.C., Rahman, M., Himelein-Wachowiak, M., Kwarteng, A., Curtis, B.: Twitter Corpus of the #BlackLivesMatter Movement And Counter Protests: 2013 to 2020 (2020). https://doi.org/10.5281/zenodo.4056563
16. Goldstein, J.S.: A conflict-cooperation scale for WEIS events data. J. Conflict Resolut. **36**(2), 369–385 (1992). https://doi.org/10.1177/0022002792036002007
17. Gonzalez, J.E., Xin, R.S., Dave, A., Crankshaw, D., Franklin, M.J., Stoica, I.: GraphX: graph processing in a distributed dataflow framework. In: Proceedings of the 11th USENIX Conference on Operating Systems Design and Implementation, OSDI 2014, pp. 599–613. USENIX Association (2014)
18. Granger, C.: Testing for causality - a personal viewpoint. J. Econ. Dyn. Control **2** (1980)
19. Hawkes, A.G.: Spectra of some self-exciting and mutually exciting point processes. Biometrika **58**, 83–90 (1971). https://www.jstor.org/stable/2334319
20. Herman, E., Chomsky, N.: Manufacturing Consent: The Political Economy of the Mass Media. Random House (2010). https://books.google.se/books?id=Kv_-bvCqgrEC
21. Jackson, S.J., Bailey, M., Foucault Welles, B.: #HashtagActivism: Networks of Race and Gender Justice. The MIT Press (2020). https://doi.org/10.7551/mitpress/10858.001.0001
22. Jansen, B.J., Zhang, M., Sobel, K., Chowdury, A.: Twitter power: tweets as electronic word of mouth. J. Am. Soc. Inf. Sci. Tech. **60**(11), 2169–2188 (2009)
23. Leetaru, K., Schrodt, P.A.: GDELT: Global data on events, location, and tone. ISA Annual Convention (2013)
24. Lindström, A., Lindgren, S., Sainudiin, R.: Hawkes processes on social and mass media: a causal study of the #blacklivesmatter movement in the summer of 2020. In: Proceedings of the 12th International Conference on Data Science, Technology and Applications - DATA, pp. 77–88. INSTICC, SciTePress (2023). https://doi.org/10.5220/0012089500003541
25. Lippmann, W.: Public opinion (1922). https://www.wikidata.org/wiki/Q1768450. Wikidata Q1768450
26. Malewicz, G., et al.: Pregel: a system for large-scale graph processing. In: Proceedings of the 2010 ACM SIGMOD International Conference on Management of Data, pp. 135–146 (2010)
27. Metaxas, P., Mustafaraj, E., Wong, K., Zeng, L., O'Keefe, M., Finn, S.: What do retweets indicate? Results from user survey and meta-review of research. In: International AAAI Conference on Web and Social Media. ACM (2015). https://www.aaai.org/ocs/index.php/ICWSM/ICWSM15/paper/view/10555
28. Mishra, S., Rizoiu, M.A., Xie, L.: Feature driven and point process approaches for popularity prediction. In: Proceedings of the 25th ACM International on Conference on Information and Knowledge Management, CIKM 2016, pp. 1069–1078. Association for Computing Machinery, New York (2016). https://doi.org/10.1145/2983323.2983812
29. Muphy, D.: 44 Percent of Twitter Accounts Have Never Tweeted. PCMag UK (2014). https://uk.pcmag.com/social-media/9422/44-percent-of-twitter-accounts-have-never-tweeted
30. Ngo, A.: Unmasked: Inside Antifa's Radical Plan to Destroy Democracy. Center Street (2021)
31. O'Reilly, T., Strauss, I., Mazzucato, M.: Algorithmic Attention Rents: a theory of digital platform market power. UCL Institute for Innovation and Public Purpose, Working Paper Series (IIPP WP 2023-10) (2023). https://www.ucl.ac.uk/bartlett/public-purpose/wp2023-10
32. Pierris, G.D., Friedman, M.: Kant and Hume on Causality. The Stanford Encyclopedia of Philosophy (Winter 2018 Edition) (2018)
33. Raghavan, U.N., Albert, R., Kumara, S.: Near linear time algorithm to detect community structures in large-scale networks. Phys. Rev. E **76** (2007). https://arxiv.org/abs/0709.2938
34. Rizoiu, M.A., Graham, T., Zhang, R., Zhang, Y., Ackland, R., Xie, L.: #debatenight: the role and influence of socialbots on Twitter during the 1st 2016 U.S. presidential debate. In:

Proceedings of the International AAAI Conference on Web and Social Media, vol. 12, no. 1 (2018). https://doi.org/10.1609/icwsm.v12i1.15029. https://ojs.aaai.org/index.php/ICWSM/article/view/15029
35. Sainudiin, R., Yogeeswaran, K., Nash, K., Sahioun, R.: Characterizing the Twitter networks of prominent politicians and SPLC-defined hate groups in the 2016 US presidential election. Soc. Netw. Anal. Min. **9**(34) (2019)
36. Santos, T., Lemmerich, F., Helic, D.: Surfacing estimation uncertainty in the decay parameters of Hawkes processes with exponential kernels. CoRR abs/2104.01029 (2021). https://arxiv.org/abs/2104.01029
37. Schrodt, P.A.: CAMEO - Conflict and Mediation Event Observations Event and Actor Codebook (2012). http://data.gdeltproject.org/documentation/CAMEO.Manual.1.1b3.pdf
38. Strittmatter, K.: We Have Been Harmonised: Life in China's Surveillance State. Old Street Publishing Limited (2019). https://books.google.se/books?id=bEzHxQEACAAJ
39. Trappel, J., Tales Tomaz, T.: Euromedia ownership monitor. https://media-ownership.eu/. Accessed 10 Nov 2023
40. Twitter: Twitter Shareholder letter, Q4 2020. https://investor.twitterinc.com/files/doc_financials/2020/q4/FINAL-Q4'20-TWTR-Shareholder-Letter.pdf
41. Zhao, Q., Erdogdu, M.A., He, H.Y., Rajaraman, A., Leskovec, J.: SEISMIC: a self-exciting point process model for predicting tweet popularity. In: Proceedings of the 21th ACM SIGKDD International Conference on Knowledge Discovery and Data Mining (2015). https://doi.org/10.1145/2783258.2783401
42. Zhou, K., Zha, H., Song, L.: Learning social infectivity in sparse low-rank networks using multi-dimensional Hawkes processes. In: Proceedings of the Sixteenth International Conference on Artificial Intelligence and Statistics, vol. 31, pp. 641–649 (2013). https://proceedings.mlr.press/v31/zhou13a.html
43. Zuboff, S.: The Age of Surveillance Capitalism: The Fight for a Human Future at the New Frontier of Power. Profile Books (2019). https://books.google.se/books?id=bYL9xAEACAAJ

A Survival Analysis Approach to Predicting Severe Leukopenia in Advanced Breast Cancer Patients

Asmir Vodenčarević[1](✉), Julia Kreuzeder[1], Achim Wöckel[2], and Peter A. Fasching[3]

[1] Novartis Pharma GmbH, Nuremberg, Germany
asmir.vodencarevic@novartis.com
[2] University Hospital Würzburg, Würzburg, Germany
[3] University Hospital Erlangen, Erlangen, Germany

Abstract. Modern therapies for advanced breast cancer can significantly prolong survival and improve patient's quality of life. Yet, in some patients they can be associated with different, sometimes severe toxicities. Predicting these toxicities in individual patients would enable more targeted monitoring, toxicity prevention and could help saving scarce healthcare resources. Previously we have shown that individual prediction of a cardio toxicity called QT prolongation using survival modelling algorithms is feasible. In this work we target the individual prediction of a blood toxicity called severe leukopenia using the same modelling approach. Seven statistical and machine learning survival modelling algorithms are trained, optimized, validated and compared on both clinical trial and real-world data. The statistical Cox proportional hazards model reached rather modest performance of about 0.6 (Uno's concordance index) and slightly outperformed its machine learning counterparts. The strongest predictors include variables related to prior therapies, location of metastasis, histologic grade as well as several quality of life scores.

Keywords: Severe leukopenia · Survival analysis · Advanced breast cancer

1 Introduction

2.3 million new cases in 2020 with a growing tendency make breast cancer the most common type of cancer in women [1]. Approximately one third of the patients diagnosed with early stage breast cancer develop incurable metastatic disease spreading to other organs like liver, brain, lungs and bones. Nonetheless, nowadays there are a number of treatments available, which could ease the symptoms and prolong survival of patients. They include radiotherapy, chemotherapy, surgery, immunotherapy, hormone and targeted therapy. These treatments as well as certain risk factors like advanced age, can be associated with different, potentially severe or even serious toxicities such as cardio or blood toxicity.

O. Gusikhin et al. (Eds.): DATA 2023, CCIS 2105, pp. 127–140, 2024.
https://doi.org/10.1007/978-3-031-68919-2_6

In our previous work [2] we have applied a survival analysis approach to predict a specific cardiac event called QT prolongation. A QT prolongation is an abnormal extension of the time interval between the start of the Q wave and the end of the T wave in a patient's electrocardiogram, which can in the worse cases lead to cardiac arrythmia and even sudden death. We have shown that prediction of QT prolongation in individual patients is feasible with modest to moderate performance [2] using both statistical and machine learning survival modelling algorithms. Moreover, clinically meaningful predictive factors have been identified. These positive results have motivated further research on predicting other important safety outcomes using survival analysis approach. One such outcome is a hematological toxicity known as leukopenia. In simple terms, leukopenia is a condition characterized by a low count of total White Blood Cells (WBC), which play a crucial role in defending human body from bacteria and viruses. According to the Common Terminology Criteria for Adverse Events (CTCAE) [3], leukopenia is categorized in 5 grades, ranging from mild (grade 1) to death (grade 5). In this work we focus on medically significant (grade 3, WBC $< 2000/mm^3$) and life-threatening (grade 4, WBC $< 1000/mm^3$) leukopenia, which are commonly referred to as *severe* leukopenia.

Conventional machine learning algorithms have been applied before to predict hematological toxicities like thrombocytopenia, neutropenia, and lymphopenia in cancer patients, e.g. suffering from glioblastoma (cancer of brain and spinal cord) [4] or pediatric renal cancer [5]. A prediction model for drug-induced severe leukopenia has been developed for non-small cell lung cancer patients [6] using data from whole-exome sequencing. However, validation performance has not been reported. Similarly, a multivariate logistic regression model was fit to identify predictors of leukopenia using data of 81 patients with castration-resistant prostate cancer [7]. Pretreatment WBC $< 5500/mm^3$ and age were reported as significant predictors for severe (grade3/4) leukopenia without reporting predictive performance of the model. In [8] a LASSO classification model was trained and optimized to predict leukopenia in a small cohort of 35 osteosarcoma (a type of bone cancer) patients. External validation in a very small cohort of 12 patients measured an Area Under the Receiver Operating Characteristic curve (AUROC) value of 0.832 (95% CI: 0.695–0.970) and the most predictive factors included gender, drug concentration in blood and blood markers Ki-67, platelet count and aspartate aminotransferase. Oyaga-Iriarte [9] applied and compared several machine learning algorithms for binary classification to predict leukopenia and other toxicities in patients suffering from metastatic colorectal cancer. Random forest model reached the highest (internal) validation AUROC of 0.74. While technically sound, this work suffers from the small sample size (20 patients with 53 cycles of chemotherapy in total) and the lack of external validation. In [10] the task of predicting severe leukopenia in advanced breast cancer patients was addressed by applying conventional binary classification algorithms LASSO and XGBoost and using data from the same, relatively large two studies as in this work (described in the next section). The best model reached the AUROC of about 0.6 when cross-validated within one study but it didn't generalize well neither in the other study nor in the cross-study validation. To the best of our knowledge, there are no published approaches aiming at prediction of severe leukopenia in advanced breast cancer patients using survival modelling algorithms.

Contributions of this work are as follows: First, we investigate the feasibility of predicting severe leukopenia in advanced breast cancer patients using baseline (prior to treatment start) data and survival modelling algorithms. As in our previous work [2], patients treated with the Cyclin-Dependent Kinase 4 and 6 (CDK4/6) inhibitor ribociclib are included in this study. Second, we comprehensively (internally and externally) validate and compare seven statistical and machine learning survival modeling algorithms using two patient cohorts much larger than in the related work. Third, we identify top predictors of severe leukopenia based on the best performing model.

2 Data

For training, optimizing and validating prediction models for severe leukopenia, data from two clinical studies were used:

- RIBECCA [11]: a national, multicenter single-arm, open-label phase 3b clinical trial investigating the efficacy and safety of the treatment with CDK4/6 inhibitor ribociclib plus letrozole in patients with Hormone Receptor–positive/Human Epidermal growth factor Receptor 2–negative (HR+/HER2-) advanced (recurrent or metastatic) breast cancer
- RIBANNA [12]: a non-interventional study which evaluates the real-world efficacy and safety of first-line CDK4/6 inhibitor ribociclib in combination with aromatase inhibitor/fulvestrant, endocrine monotherapy or chemotherapy.

Detailed descriptions of these studies can be found in the corresponding references. All data was anonymized prior to processing.

2.1 Data Selection

Patient Selection. An overview of the patient selection process is depicted in Fig. 1. Patients whose data was recoded at baseline (prior to treatment start) were considered for the analysis. Inclusion criterion relates to patients who received at least one dose of study medication. Patients not treated with ribociclib were excluded from analysis as well as patients with unrealistic data points. The latter included e.g. patients having at least one non-positive electrocardiogram interval (e.g. non-positive time interval between the start of the P wave and the end of the R wave), having zero blood pressure or negative time since primary diagnosis. Compared to RIBECCA significantly more patients from RIBANNA, which contains real-world data with correspondingly lower data quality, had to be excluded.

Variable Selection. As in any supervised modelling task, input and target variables are defined prior to data preparation. Both RIBECCA and RIBANNA studies recorded about 420 variables. The majority of them were not relevant for our modelling task, as they related to various absolute dates, follow-up visits and placeholders for tumor control and safety variables. Furthermore, only variables which were recorded in both studies were selected. This criterion was necessary in order to be able to perform a cross-study (external) validation of prediction models. Finally, 72 variables recorded at baseline and potentially relevant for predicting severe leukopenia were included into the analysis as input variables. They can be grouped as follows:

- Demographic data like age and body-mass index
- Diagnosis and cancer severity variables like time since primary diagnosis, histologic grade and metastasis location (e.g. brain, lung, bones, liver)
- Prior therapies like surgery and radiotherapy
- Hormone receptor status including estrogen and progesterone
- Vital signs like electrocardiogram features, systolic and diastolic blood pressure
- Medical history like fatigue and pneumonia
- Measurements of patient's everyday cognitive abilities from the ECOG (Eastern Cooperative Oncology Group) status scale and patient reported outcomes from different EORTC (European Organization for Research and Treatment of Cancer) quality of life questionnaires.

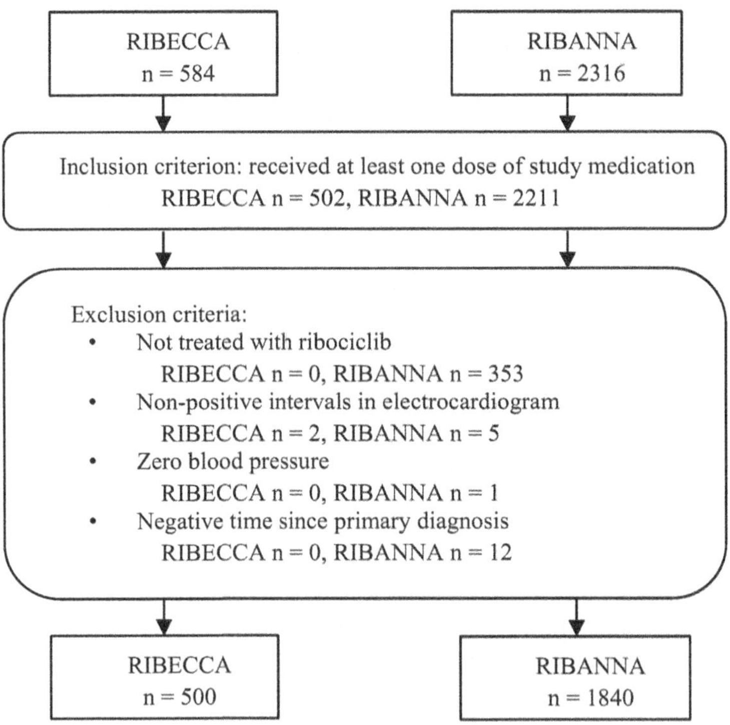

Fig. 1. Overview of the patient selection process.

The target variable was severe leukopenia, defined as either medically significant (grade 3, WBC < 2000/mm^3) or life-threatening (grade 4, WBC < 1000/mm^3) leukopenia. Similar to the adverse event QT prolongation modelled in our previous work [2], severe leukopenia was also recorded as a binary event indicator $e \in \{0, 1\}$ together with the event absolute date, in case it happened. For patients who discontinued from the study before it ended (e.g. lost to follow-up) and those who did not experience an event until the end of the study, the target variable was censored. It remains unknown if the event has happened after the censoring date, i.e. after last contact with the patient. This setting makes it easy to transform the problem of predicting severe leukopenia into a survival analysis problem by computing the time of event if the event has occurred and the time of censoring if the event hasn't occurred, both relative to the baseline date. The computation is trivial as baseline, event and censoring absolute dates are recorded for each patient. In mathematical terms, the observable time y for a patient is defined as follows:

$$y = \min(t, c) = \begin{cases} t \; if \;\; e = 1, \\ c \; if \;\; e = 0. \end{cases} \tag{1}$$

where $t > 0$ is the event time and $c > 0$ is the censoring time. Finally, the target variable was constructed as a structured array of (e, y) pairs.

Just like with QT prolongation [2] and as typical for survival analysis problems, severe leukopenia target variable was imbalanced as well. It was recorded in 82 (16.4%) patients from the RIBECCA study and 99 (5.4%) patients from the RIBANNA study, with median observation times of 16.5 and 30 days, respectively. The estimated probability of severe leukopenia at certain time intervals is shown in Kaplan-Meier curves (with 95% confidence intervals shaded) depicted in Fig. 2 and Fig. 3 for RIBECCA and RIBANNA, respectively. It is interesting to note that the majority of events occurred relatively shortly after start of the studies. About 88% (72 from 82) of the events in RIBECCA and about 94% (93 from 99) of the events in RIBANNA occurred already within the first 200 days.

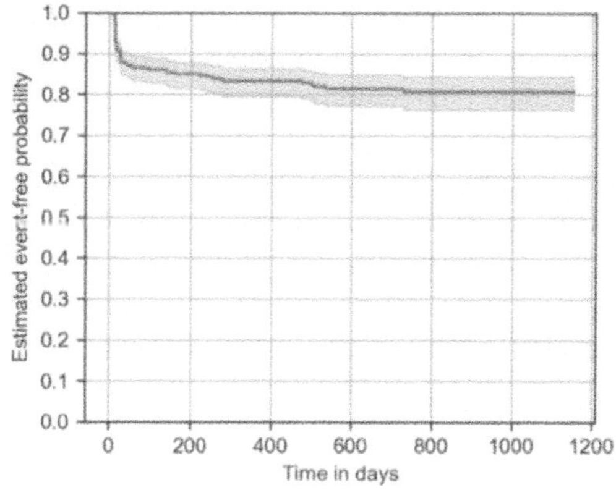

At risk	500	263	197	146	90	36	0
Censored	0	165	226	273	328	382	418
Events	0	72	77	81	82	82	82

Fig. 2. Kaplan-Meier curve for severe leukopenia in RIBECCA.

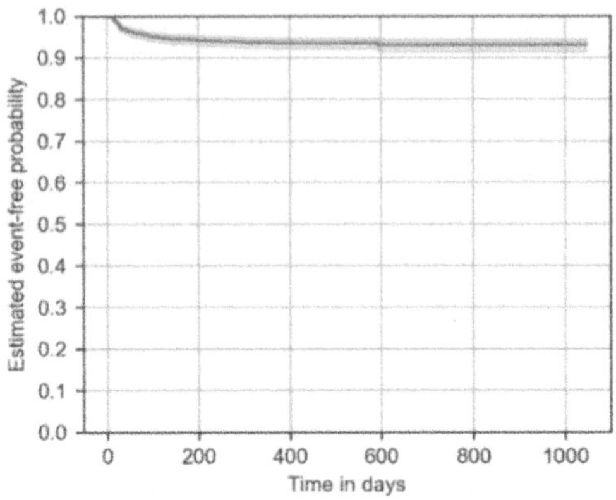

At risk	1840	929	498	202	64	5
Censored	0	818	1244	1539	1677	1736
Events	0	93	98	99	99	99

Fig. 3. Kaplan-Meier curve for severe leukopenia in RIBANNA.

2.2 Data Preparation

To avoid any potential data leakage, data preparation prior to modelling was done in an unsupervised fashion, not involving the target variable. Data recorded in the closely monitored RIBECCA clinical trial had significantly higher quality than the real-world data from the non-interventional study RIBANNA. After the initial patient and variable selection described in Sect. 2.1, RIBECCA and RIBANNA datasets still contained 4.6% and 24.5% of missing values in the baseline input data, respectively.

Basically, the same data preparation steps as in our previous work [2] have been performed:

- Four variables with more than 50% of missing values in either study were removed
- Pearson correlation coefficient for numerical variables was checked to identify redundant variables. Patient weight and height were removed and body-mass index was kept in the dataset
- 17 binary variables with a low frequency level in less than 1% of data were removed (as constant or close to constant variables)
- Two categorical variables with high Cramer's V association (>0.8) with other variables were removed

After removal of these 25 variables, the final dataset consisted of 32 numerical and 15 categorical variables. The summary statistics for several selected important variables are given in Table 1.

Table 1. Baseline characteristics of RIBECCA and RIBANNA patients. EORTC = European Organization for Research and Treatment of Cancer, ECOG = Eastern Cooperative Oncology Group.

Variable	RIBECCA		RIBANNA	
	Mean (std) or n (%)	Missing (%)	Mean (std) or n (%)	Missing (%)
Age (years)	63.8 (11.6)	0 (0)	64.3 (11.6)	0 (0)
Body-mass index (kg/m^2)	26.5 (4.9)	2 (0.4)	27.1 (5.7)	145 (7.9)
EORTC physical functioning revised [0,100]	26.9 (23.5)	28 (5.6)	35.5 (26.5)	433 (23.5)
EORTC pain [0,100]	36.6 (33.1)	30 (6)	41.3 (34.4)	433 (23.5)
EORTC fatigue [0,100]	40.9 (28.3)	28 (5.6)	49.1 (29.4)	432 (23.5)
ECOG class		10 (2)		256 (13.9)
0	328 (65.6)		783 (42.6)	
1	146 (29.2)		652 (35.4)	
≥ 2	16 (3.2)		149 (8.1)	
Prior surgery		0 (0)		6 (0.3)
yes	449 (89.8)		1147 (62.4)	
no	51 (10.2)		687 (37.3)	
Prior radiotherapy		0 (0)		5 (0.3)
yes	339 (67.8)		969 (52.7)	
no	161 (32.2)		866 (47)	
Locally advanced		0 (0)		21 (1.1)
yes	143 (28.6)		1411 (76.7)	
no	357 (71.4)		408 (22.2)	
Metastatic bone		0 (0)		89 (4.8)
yes	356 (71.2)		1170 (63.6)	
no	144 (28.8)		581 (31.6)	
Metastatic lung		0 (0)		89 (4.8)
yes	140 (28)		441 (24)	
no	360 (72)		1310(71.2)	

3 Methods

We modelled severe leukopenia using seven statistical and machine learning survival modelling algorithms. In this section, the employed algorithms are briefly described together with a machine learning pipeline for algorithm optimization and validation. The later includes descriptions of the selected performance metric and feature importance approach applied to identify the most important predictive factors.

3.1 Algorithms

Predictive models for severe leukopenia were trained using the following algorithms: classical statistical Cox Proportional Hazards model (CPH), Cox Proportional Hazards model penalized by elastic Net (CPHNet), accelerated failure time model with Inverse

Probability of Censoring weights (IPC), Gradient Boosting Survival algorithm (GBS), Random Survival Forest (RSF), Extra Survival Trees (EST) and fast Survival Support Vector Machines (SSVM).

CPH. CPH is a well-known statistical regression algorithm for survival analysis used to estimate the risk of an event over time and to identify predictive factors. Its major strengths are its simplicity and interpretability (so called "white box" model), which made it popular, especially in medical applications. However, it typically does not perform well in problems with high-dimensional input data, and it is ineffective in modelling, non-linearities, interaction effects and correlated variables. Regularization with the L2 penalty can help with the latter issue.

CPHNet. CPHNet extends CPH by implementing the elastic net regularization, which makes a trade-off between the L2 and L1 penalties. In this way, numerical stability of the algorithm is enhanced, making it more suitable for modelling high-dimensional data and correlated variables. It does not address other issues of CPH.

IPC. Another statistical algorithm which uses L2 penalty to address correlated variables is called IPC. It is especially suited for modelling right censored data as it weights each data point (i.e. a patient dataset) by the inverse probability of censoring. In this way, unobservable data points have smaller impact on the computation of the model's coefficients.

GBS. A (so called "black box") machine learning algorithm for survival analysis based on the conventional gradient boosting algorithm is GBS. It trains a sequence of simple models called "weak" learners like shallow decision trees, where each next learner in a sequence assigns higher weight to the data points predicted wrongly by previous learners. Each tree tries to group data points (i.e. patients) based on the event indicator and its observed time. GBS belongs to ensemble algorithms that can model non-linear relationships and interactions between variables.

RSF. RSF is another ensemble, machine learning, survival modelling algorithm. It is also based on its conventional counterpart: the random forest algorithm. In contrast to GBS, it trains decision trees in parallel on randomly selected bootstrap replicas of data, created by sampling with replacement. During the training process, tree nodes are split based on the most discriminative thresholds, which can be determined in different ways. The final prediction is computed by aggregating predictions of each tree. Like GBS, it is also suitable for modelling non-linearities and interactions.

EST. An algorithm very similar to RSF with additional level of randomness introduced is called EST. In EST, tree nodes are split using the best threshold from a set of randomly drawn thresholds. Furthermore, by default EST uses the original data for training each tree and typically executes faster than RSF.

SSVM. A survival modelling implementation of the machine learning algorithm support vector machines is SSVM. The algorithm uses so-called kernel functions to transform the original data space into a high-dimensional space where discrimination between data points can be achieved easier. A hyperplane is fitted in a high-dimensional space, which maximizes a margin between data points with dissimilar survival (i.e. time of severe leukopenia). Data points closest to the hyperplane are called support vectors.

A detailed user guide and references for these algorithms can be found in the documentation of the Python package scikit-survival [13], which is used in our analysis. The ownership of the project source code stays with the industrial project partner.

3.2 Training, Optimization and Validation Pipeline

Performance Metric. In the process of training, optimization and validation, predictive models are compared based on an adequate performance metric. The conventional performance metric in survival analysis is Harrell's concordance index (or shorter c-index), which measures how well the model orders data points (i.e. patients) based on the event time. Its interpretation is very similar as the interpretation of the area under the receiver operating characteristic curve, often used to quantify the performance of binary classification algorithms. For a random pair of patients, Harrell's c-index is the probability that a patient with longer time to event (e.g. longer survival) will have lower predicted risk score. Harrell's c-index of 0.5 means that the model is not better than random guessing, while the value of 1 means that the model does perfect ordering of patients.

As indicated by Uno [14] and explained in our previous work [2], Harrell's c-index is inappropriate performance metric for data with extensive censoring, as it produces falsely optimistic results. Since over 80% of datasets in both RIBECCA and RIBANNA are right censored, we decided to use Uno's c-index, which mitigates the problem by introducing inverse probability of censoring weights to the computation of Harrell's c-index. Data points more likely to be observed are assigned higher weights, making the performance estimate unbiased.

Machine Learning Pipeline. After data preparation steps described in Sect. 2.2 were performed, the data still included many missing values. Numerical variables had different ranges (e.g. days since primary diagnosis in thousands, age in dozens), which is problematic for several algorithms. Moreover, categorical variables contained string values (e.g. ECOG variable had values like "$\geq$2"), which had to be encoded to numerical, binary values (0/1) in order to be processable by the algorithms. We applied the best practice for avoiding data leakage and addressed these issues by data transformations within a machine learning pipeline, which was afterwards provided to optimization and validation procedures. More concretely, machine learning pipeline included:

- Iterative imputer based on Bayesian ridge regression algorithm [15] for imputing missing values in numerical variables
- Min-max normalization for scaling ranges of numerical variables
- Simple imputer based on the most frequent value for imputing missing values in categorical variables
- One-hot encoder for creating binary variables for each category of categorical variables
- Survival modelling algorithm which was optimized and validated.

Optimization. No free lunch theorem [16] states that any two algorithms are equivalent when their performance is averaged over all possible problems. Therefore, it could be beneficial to test different algorithms and their hyperparameter values in each specific problem. This is the rationale for deciding to test seven algorithms in our work. Moreover, we decided to optimize specific algorithm hyperparameters for higher predictive performance as measured by Uno's c-index. A set of hyperparameter values was predefined for every algorithm, and they were evaluated in a grid search procedure using 2-fold cross-validation. Table 2 gives an overview of optimized hyperparameters for each algorithm.

Table 2. Overview of algorithm hyperparameters that were optimized.

Algorithm	Hyperparameters
CPH	Strength of the L2 regularization penalty
CPHNet	Elastic net ratio between L1 and L2 regularization
IPC	Strength of the L2 regularization penalty
GBS	Learning rate Maximum tree depth
RSF	Minimum samples in leaf nodes Maximum tree depth
EST	Minimum samples in leaf nodes Maximum tree depth
SSVM	Strength of the L2 regularization penalty

Internal Validation. Predictive performance of models trained and optimized within machine learning pipelines as described above have first been validated internally, i.e. separately within RIBECCA and RIBANNA. To avoid overfitting and obtain unbiased estimates of the real model performance while optimizing algorithm hyperparameters, we implemented a 3 × 2-fold nested cross-validation [17]. The inner, 2-fold cross-validation loop searched for the best hyperparameter values, while the outer, 3-fold cross-validation measured the model performance.

External Validation. The goal of the external validation was to train models on data from one study and measure their performance on data from another study. To this end, the outer 3-fold cross-validation loop was excluded. All data from one study was provided to the 2-fold cross-validation which performed algorithm optimization. The best model was then simply applied to the data from another study to measure predictive performance. We performed this in a cross-study fashion, meaning that the training and validation studies were then switched and the experiment was repeated.

3.3 Feature Importance

Feature permutation importance method [18] was used to identify predictors of severe leukopenia. The method measures decrease in the model performance, when a single variable is randomly shuffled, mimicking variable exclusion from the analysis. Those variables whose random shuffling causes the biggest drop in the model performance are considered strongest predictors. Feature permutation importance was applied to the best models obtained in the external, cross-study validation (trained first on RIBECCA and then on RIBANNA) and the results were compared.

4 Results

4.1 Performance Comparison

As described in Sect. 3, seven applied models were trained, optimized and validated both internally, using the 3 × 2-fold nested cross-validation procedure, as well as externally, in a cross-study fashion. The predictive performance of models given as Uno's c-index are shown in Table 3.

The overall model performance was rather modest. The highest and the most consistent values of Uno's c-index were demonstrated by a classical statistical CPH model. More flexible machine learning models underperformed in at least one validation. Only EST showed stable performance across all validations, which came close to the performance of CPH. The modest predictive performance could at least partially be attributed to the lack of blood markers in RIBECCA and RIBANNA data, which were identified as strong predictors of leukopenia in related work.

Table 3. Uno's c-index measured in internal (as mean (std)) and external validations.

Algorithm	Internal within RIBECCA	Internal within RIBANNA	Trained on RIBECCA, validated on RIBANNA	Trained on RIBANNA, validated on RIBECCA
CPH	0.60 (0.04)	0.63 (0.04)	0.60	0.62
CPHNet	0.60 (0.06)	0.62 (0.05)	0.56	0.46
IPC	0.47 (0.07)	0.61 (0.03)	0.51	0.55
GBS	0.60 (0.05)	0.61 (0.06)	0.58	0.57
RSF	0.60 (0.04)	0.57 (0.08)	0.57	0.56
EST	0.59 (0.05)	0.60 (0.06)	0.59	0.59
SSVM	0.57 (0.05)	0.60 (0.07)	0.55	0.62

4.2 Predictors

As described in Sect. 3.3, strongest predictions of severe leukopenia were identified by applying the feature permutation importance method to the best and most consistent model, which was CPH. This was done for both the CPH model trained on RIBANNA and validated on RIBECCA and vice versa. The results are shown in Fig. 4 and Fig. 5, respectively.

The strongest predictor was prior radiotherapy in both evaluations. Under the top ten predictors, seven were common for both CPH models and in addition to prior radiotherapy, they include cancer severity features (histologic grade and metastases in lung and bone) and EORTC quality of life scores for physical functioning, fatigue and pain. In general, prior therapies, cancer severity and quality of life variables are dominant predictors of severe leukopenia in both evaluations.

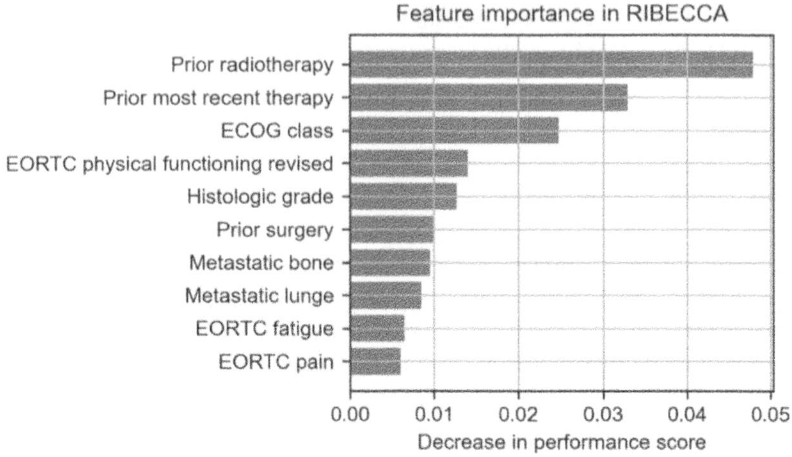

Fig. 4. Identified top predictors for the model trained on RIBANNA and validated on RIBECCA.

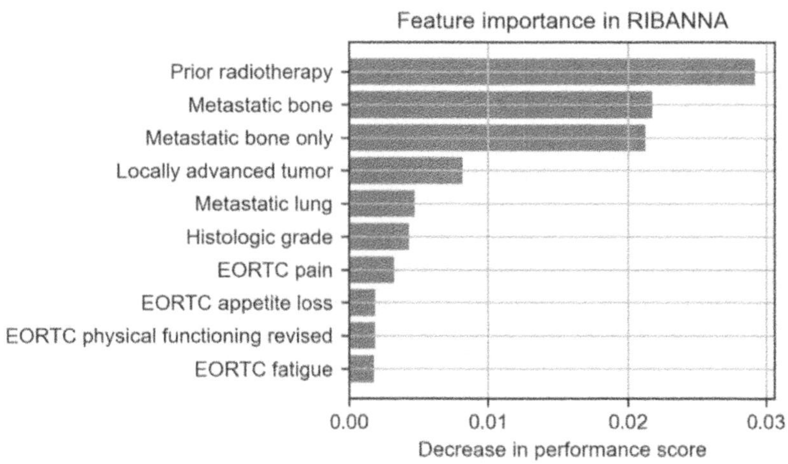

Fig. 5. Identified top predictors for the model trained on RIBECCA and validated on RIBANNA.

5 Conclusions

In this work we investigated feasibility of predicting severe leukopenia after starting ribociclib treatment using survival modelling algorithms and data from the clinical trial RIBECCA and the non-interventional study RIBANNA. Seven statistical and machine learning survival modelling algorithms were trained, optimized, validated both internally and externally. The use of common performance metric (Uno's c-index) enabled direct performance comparison of models trained with these algorithms. Only a modest performance was reached, which can be partially attributed to the lack of blood markers such as platelet count, aspartate aminotransferase or pretreatment count of white blood cells, which were reported as significant in the related work. The best model was statistical Cox proportional hazards model with consistent performance of at least 0.6 across all validations. Prior radiotherapy was identified as the strongest predictor of severe leukopenia in our both datasets and other important predictors include cancer severity variables (histologic grade and metastasis locations) and EORTC quality of life scores. In general, this and our previous work [2] testify about the utility of survival analysis in modelling tumor safety outcomes and beyond.

In the future work we will apply survival modelling algorithms to the problem of predicting tumor control outcomes such as progression-free survival. Moreover, we will apply SHAP explainable artificial intelligence approach [19] to generate more insights regarding predictors and enable explainability of individual, patient-level predictions.

References

1. Arnold, M., Morgan, E., Rumgay, H., Mafra, A., Singh, D., Laversanne, M., et al.: Current and future burden of breast cancer: global statistics for 2020 and 2040. Breast **66**, 15–23 (2022). https://doi.org/10.1016/j.breast.2022.08.010
2. Vodenčarević, A., Kreuzeder, J., Wöckel, A., Fasching, P.A.: Prediction of QT prolongation in advanced breast cancer patients using survival modelling algorithms. In: Proceedings of the 12th International Conference on Data Science, Technology and Applications DATA, vol. 1, pp. 164–172. SciTePress, Rome, Italy (2023). https://doi.org/10.5220/0012130900003541
3. Common Terminology Criteria for Adverse Events (CTCAE) Homepage, https://ctep.cancer.gov/protocoldevelopment/electronic_applications/ctc.htm, v5.0 published 2017/11/27
4. Amouheidari, A., Alirezaei, Z., Rauh, S., Hassanpour, M.: PrACTiC: a predictive algorithm for chemoradiotherapy-induced cytopenia in glioblastoma patients. J. Oncol. **2022**, 1438190 (2022). https://doi.org/10.1155/2022/1438190
5. Li, M., Wang, Q., Lu, P., Zhang, D., Hua, Y., Liu, F., et al.: Development of a machine learning-based prediction model for chemotherapy-induced myelosuppression in children with Wilms' tumor. Cancers **15**(4), 1078 (2023). https://doi.org/10.3390/cancers15041078
6. Svedberg, A., Sigurgeirsson, B., Björn, N., Pradhananga, S., Brandén, E., Koyi, H., et al.: Association to drug-induced leukopenia using whole-exome sequencing of non-small cell lung cancer patients on gemcitabine/carboplatin regimen. Cancer Res. **77**(13_Supplement), 5026 (2017). https://doi.org/10.1158/1538-7445.AM2017-5026
7. Nakashima, J., Takizawa, I., Ko, K., Nakagami, Y., Ohori, M., Ohno, Y., et al.: Prediction of leukopenia after docetaxel chemotherapy in patients with castration-resistant prostate cancer. J. Urol. **187**(4S), e311 (2012). https://doi.org/10.1016/j.juro.2012.02.848
8. Wu, H., Xu, G., Li, Z., Xu, Y., Lin, Y., Chekhonin, V.P., et al.: Nomogram predicting leukopenia in osteosarcoma after high-dose methotrexate chemotherapy. Aging **14**(12), 5023–5033 (2022). https://doi.org/10.18632/aging.203978
9. Oyaga-Iriarte, E., Insausti, A., Sayar, O., Aldaz, A.: Prediction of irinotecan toxicity in metastatic colorectal cancer patients based on machine learning models with pharmacokinetic parameters. J. Pharmacol. Sci. **140**(1), 20–25 (2019). https://doi.org/10.1016/j.jphs.2019.03.004
10. Fasching, P.A., Wöckel, A., Tesch, H., Volz, B., Pritzsche, U., Bachmann M., et al.: Machine learning to predict treatment response and tolerability in HR+, HER2– advanced breast cancer: German study AI4ANNA. Cancer Res. **83**(5_Supplement), P4–01–05 (2023). https://doi.org/10.1158/1538-7445.SABCS22-P4-01-05

11. Decker, T., Fasching, P.A., Nusch, A., Hartkopf, A., Heinrich, B.J., Kurbacheret, C.M., et al.: Efficacy and safety of ribociclib (RIB) in combination with letrozole (LET) in patients with estrogen receptor–positive advanced breast cancer (ABC): secondary and exploratory results of phase 3b RIBECCA study. Ann. Oncol. **32**(suppl_5), S457–S515 (2021). https://doi.org/10.1016/annonc/annonc689
12. Lüftner, D., Brucker, C., Decker, T., Fasching, P.A., Göhler, T., Jackisch, C., et al.: Real-world efficacy of ribociclib (RIB) plus aromatase inhibitor (AI)/fulvestrant (FUL), or endocrine monotherapy (ET), or chemotherapy (CT) as first-line (1L) treatment (tx) in patients (pts) with hormone receptor–positive (HR+), human epidermal growth factor receptor-2–negative (HER2–) advanced breast cancer (ABC): results of fourth interim analysis (IA) from RIBANNA. J. Clin. Oncol. **40**(16_suppl), 1065 (2022). https://doi.org/10.1200/JCO.2022.40.16_suppl.1065
13. Pölsterl, S.: Scikit-survival: a Library for time-to-event analysis built on top of scikit-learn. J. Mach. Learn. Res. **21**(212), 8747–8752 (2020). https://doi.org/10.5555/3455716.3455928
14. Uno, H., Cai, T., Pencina, M.J., D'Agostino, R.B., Wei, L.J.: On the C-statistics for evaluating overall adequacy of risk prediction procedures with censored survival data. Stat. Med. **30**(10), 1105–1117 (2011). https://doi.org/10.1002/sim.4154
15. Bishop, C.: Pattern Recognition and Machine Learning, 1st edn. Springer-Verlag, New York (2006)
16. Wolpert, D.H., Macready, W.G.: No free lunch theorems for optimization. IEEE Trans. Evol. Comput. **1**(1), 67–82 (1997). https://doi.org/10.1109/4235.585893
17. Cawley, G.C., Talbot, N.L.C.: On over-fitting in model selection and subsequent selection bias in performance evaluation. J. Mach. Learn. Res. **11**, 2079–2107 (2010). https://doi.org/10.5555/1756006.1859921
18. Breiman, L.: Random forests. Mach. Learn. **45**(1), 5–32 (2001). https://doi.org/10.1023/A:1010933404324
19. Lundberg, S.M., Lee, S.I.: A unified approach to interpreting model predictions. In: Proceedings of the 31st International Conference on Neural Information Processing Systems, pp. 4768–4777 (2017). https://doi.org/10.48550/arXiv.1705.07874

Author Index

O. Gusikhin et al. (Eds.): DATA 2023, CCIS 2105, p. 141, 2024.
https://doi.org/10.1007/978-3-031-68919-2

The manufacturer's authorised representative in the EU is Springer Nature Customer Service Centre GmbH, Europaplatz 3, 69115 Heidelberg, Germany. If you have any concerns regarding our products, please contact ProductSafety@springernature.com

Printed and bound by CPI Group (UK) Ltd, Croydon, CR0 4YY
30/09/2026
02225576-0009